Optical and Wireless Communications

NEXT GENERATION NETWORKS

Optical and Wireless Communications

NEXT GENERATION NETWORKS

MATTHEW N. O. SADIKU

Boeing Satellite Systems
Los Angeles, California

CRC PRESS

Boca Raton London New York Washington, D.C.

Cover Art: Image adapted from *IEEE Communications Magazine*, Summer 2001, Vol. 39, No. 9. (© 2001 IEEE. With permission.)

Library of Congress Cataloging-in-Publication Data

Sadiku, Matthew N. O.

Optical and wireless communications : next generation networks / Matthew N.O. Sadiku.

p. cm.

Includes bibliographical references and index.

ISBN 0-8493-1278-7 (alk. paper)

1. Optical communications. 2. Wireless communication systems. I. Title.

TK5103.59 .S33 2002

621.382′7--dc21 2001052889

Direct all inquiries to CRC Press LLC, 2000 N.W. Corporate Blvd., Boca Raton, Florida 33431.

Visit the CRC Press Web site at www.crcpress.com

International Standard Book Number 0-8493-1278-7

Library of Congress Card Number 2001052889

Printed in the United States of America 2 3 4 5 6 7 8 9 0

Printed on acid-free paper

Dedication

To my friends

Joshua Efuwape, Chinyere Totty, and Christopher Falade

Preface

In recent years, several next generation technologies (such as next generation computers, next generation networks, next generation web, next generation databases, next generation software infrastructure, next generation Internet) have emerged. The next generation networks are basically of two types: wired (optical fiber or copper) and wireless. This book provides a brief overview of both optical and wireless communication networks.

Optical technologies are being introduced to the global communications infrastructure at an astonishing pace, causing them to dominate the future of the communication industry. Optical networks that go to entirely new lengths are being developed. These networks are being equipped with ultra-long-reach technology and fiber-optic lines that can carry information thousands of kilometers before any signal regeneration is required.

Wireless networks offer network access virtually anytime, anyplace, and at any speed. They provide you with the connectivity of your desktop when you are on the go — you can make and receive calls, access calendars, pull up a stock quote, read headlines, send and receive e-mails. They provide the perfect solution for mobile workers who need access to information.

This book was conceived while the author was at Temple University, Philadelphia. The curriculum of the Department of Electrical and Computer Engineering at Temple University offers only two courses in data communications at the graduate level. Those two courses do not cover optical and wireless networks. This book is meant to be used in a third course to fill that gap. It is designed for a one-semester course for senior-year undergraduate and graduate engineering students. The prerequisite for the course is background knowledge of data communication in general. The book can also be used in seminars on optical and wireless networks. It can serve as a fingertip reference for engineers developing optical and wireless networks, managers involved in systems planning, and communication networks researchers and instructors.

The book is divided into two parts. Part I, on optical networks, covers optical fibers, transmitters, receivers, multiplexers, amplifiers, and specific networks such as FDDI, SONET, fiber channel, and wavelength-routed networks. Part II, on wireless networks, deals with fundamental concepts, specific wireless networks such as wireless LAN, wireless ATM, wireless local loop, and wireless PBXs, cellular technologies, and satellite communications.

I am indebted to Nora Konopka and other CRC Press staff for support. Special thanks are due to Dr. Mary Akinyemi for her encouragement and constant desire to know the progress of the book. I want to express my sincere appreciation to my colleagues at Lucent/Avaya, particularly Dr. Cyril Orji, for their help. Grateful acknowledgment is also expressed to colleagues, particularly Professor Mohammad Ilyas of Florida Atlantic University, whose reviews and constructive criticisms have improved the quality of this work. I am also grateful to my wife, Chris, and our daughters, Ann and Joyce, for their support and help.

Matthew N. O. Sadiku

About the Author

Matthew N. O. Sadiku, Ph.D., received his B.Sc. degree in 1978 from Ahmadu Bello University, Zaria, Nigeria and his M.Sc. and Ph.D. degrees from Tennessee Technological University, Cookeville, Tennessee in 1982 and 1984, respectively. From 1984 to 1988, he was an assistant professor at Florida Atlantic University, Boca Raton, where he did graduate work in computer science. From 1988 to 2000, he was at Temple University, Philadelphia, where he became a full professor. From 2000 to 2001, he worked for Lucent/Avaya, Holmdel, New Jersey, as a system engineer. Since July 2001, he has been a senior scientist with Boeing Satellite Systems, Los Angeles.

Dr. Sadiku is the author of more than 100 professional papers and 20 books, including *Numerical Techniques in Electromagnetics, Second Edition* (2001), *Metropolitan Area Networks* (1995), and with M. Ilyas, *Simulation of Local Area Networks* (1995), all published by CRC Press. Some of his books have been translated into Korean, Chinese, Italian, and Spanish. He was the recipient of the 2000 McGraw-Hill/Jacob Millman Award for outstanding contributions in the field of electrical engineering.

Dr. Sadiku's current research interests are in the areas of numerical modeling of electromagnetic systems and computer communication networks. He is a registered professional engineer and a member of the Institute of Electrical and Electronics Engineers (IEEE). He was the IEEE Region 2 Student Activities Committee Chairman. He has served as an associate editor for *IEEE Transactions on Education*.

About the Author

Introduction

My interest is in the future because I am going to spend the rest of my life there.

— Charles F. Kettering

The changing demands on communications networks is leading to the evolution of next generation networks. As the Internet becomes increasingly popular, geographic boundaries become meaningless. With a mouse click, an Internet user can trigger an information flow between nodes that are thousands of kilometers apart. With this situation, data must travel ultra-long distances and provide instantaneous connections. Next generation networks are expected to be better equipped than traditional networks to handle the traffic demands that are anticipated from tomorrow's Internet-driven economy.

The next generation networks are basically of two types: wired (optical fiber or copper) and wireless. Wireless networks should be considered extensions of the wired networks. This book is divided into two parts: Part I, optical networks, and Part II, wireless networks.

Optical networks

Optical networks are high-capacity communications networks based on optical technologies and components that provide routing, grooming, and restoration at the wavelength level as well as wavelength-based services. Several factors are driving the need for optical networks, some of which are listed below:

- *Fiber capacity*: The explosive demand for bandwidth in data networking applications continues to drive optical technology to ever-increasing capacity and flexibility at the optical physical layer. By transmitting each signal at a different frequency, network providers can send many signals on one fiber as though each is traveling on its own fiber.
- *Restoration capability*: When network planners use more network elements to increase fiber capacity, a fiber cut can have massive implications. By performing restoration in the optical layer rather than the

electrical layer, optical networks can perform protection switching more quickly and more economically.

- *Reduced cost*: With an optical network, only those wavelengths that add or drop traffic at a site need corresponding electrical nodes. Other channels can simply pass through optically, which provides tremendous cost savings in equipment and network management. Optical networks provide higher capacity and reduced costs for new applications such as the Internet, video and multimedia interaction, and advanced digital services.
- *Wavelength services*: One of the great revenue-generating aspects of optical networks is the ability to resell bandwidth rather than fiber. By maximizing the capacity available on a fiber, service providers can improve revenue by selling wavelengths, regardless of the data rate required. To customers, this service provides the same bandwidth as a dedicated fiber.

Today, transmission equipment takes electrical signals generated from communication devices, such as phones and computers, and converts them to optical signals or lightwaves that are transmitted along a strand of fiber. As the lightwave or optical signal travels over the fiber, it degrades. Once it does, equipment in the network must regenerate the lightwave by converting it into an electrical signal and then back into a lightwave before it can continue through the network. Typically, this equipment, called repeaters or regenerators, are placed every 300 to 600 km (186 to 373 mi) along the network to strengthen the lightwaves — a costly process. In an all-optical network, lightwaves can travel ultra-long distances — up to 3200 km or almost 2000 mi without being regenerated. All-optical networking is a revolutionary technology designed to change fundamentally the way data traffic is transmitted.

Wireless networks

The wireless communication revolution is bringing fundamental changes to data networking and telecommunication, and it is making integrated networks a reality. By freeing the user from the cord, personal communications networks, wireless LANs, mobile radio networks, and cellular systems promise fully distributed mobile computing and communications — anytime, anywhere. Numerous wireless services are also maturing and are positioned to change the means and scope of communication.

Wireless communication involves the biggest paradigm shift in communications. The optical-fiber revolution is about bandwidth and information distribution. The wireless revolution is more about freedom of mobility and convenience. The market for wireless communications infrastructure continues to grow at a rapid rate. Factors that contribute to this growth include:[1]

- *Insatiable appetite for communication services*: Voice is still the "killer" application for wireless communications. Despite the ability to check news, send and receive e-mails, or order things from their cell phones, consumers overwhelmingly use them to talk to other people.
- *Popularity of the Internet*: The Internet continues to expand exponentially because of worldwide accessibility. Internet technologies and applications have grown more rapidly than anyone could have envisioned even 5 years ago.
- *Deregulation in telecommunications markets*: Throughout the telecommunications marketplace, a trend toward deregulation and liberalization prompted expectations of increased competition, reduced consumer prices, and innovative new services.
- *Migration from analog to digital networks*: Early communications systems used analog transmission. Over the past two decades, digital systems have penetrated every segment of the network. A digital approach enjoys greater accuracy and stability over an analog system. The driving forces for digitization are VLSI (very large scale integrated) circuits and computers. The next generation networks will certainly employ digital transmission techniques because they can be better for most systems. The future clearly lies with digital techniques.
- *Increased competition among service providers*: The emergence of standards and continued market growth will continue to diminish the barriers to entering the market and raise the level of competition among equipment vendors.

These forces have a combined effect resulting in the need for next-generation wireless infrastructure products.

Wireless systems can be compared with wired systems.* First, the channel capacity of wireless systems typically available is much lower than that available in wired networks due to the limited spectrum available, power restrictions, and noise levels. Second, noise and interference have more impact on systems design for wireless systems than on wired systems. Third, before building a wireless system, a frequency allocation (by the Federal Communications Commission in the U.S.) is necessary. Fourth, security is a greater concern in wireless systems than in wired systems since the information may be traveling in free space.

Next generation networks

We expect that next generation networks (NGNs) will be similar to today's Internet, eventually will be as ubiquitous as the traditional telephone circuit-

* The two systems are converging in many hybrid technologies that take advantage of both. The free space optics (FSO) network is an example.

switched network, and will have performance that far exceeds that of today's fastest enterprise networks. However, they will not be simply an extension of today's Internet; we expect them to be completely different and much bigger, and to have a far greater impact on our lives.

NGNs promise to be a cost-effective way to integrate and manage voice and data over a single network. This new network infrastructure opens new avenues to do the following:[2]

- Earn greater revenue and profit through the deployment of flexible high-value service bundles
- Provide unique, integrated applications, services, and technologies that differentiate competitors
- Offer high-speed transport and switching of voice, fax, data, and video in an integrated, packet-based manner
- Achieve capital and operating savings through voice, video, and data service integration
- Unleash service innovation through industry standards and open interfaces

NGNs will be more efficient to build and less costly to operate. They will provide high value to the customer by enabling flexible, high-value service bundles such as Internet access, voice, and cable TV. The ultimate objective of NGNs is to enable one to communicate instantly with anyone else from anywhere. They will be larger than the world's telephone network and more revolutionary than the Internet and have the potential to create more wealth than did the PC industry.

References

1. G. E. Fry et al., Next generation wireless networks, *Bell Labs Tech. J.*, Autumn, 1996, 88–96.
2. D. C. Dowden, R. D. Gitlin, and R. L. Martin, Next-generation networks, *Bell Labs Tech. J.*, vol. 3, no. 4, 1998, 3–14.

Contents

part 1

Optical networks

chapter one

Optical fibers

There is one thing stronger than all the armies in the world, and that is an idea whose time has come.

— Victor Hugo

Within a short period of time, the volume of data traffic transported across communications networks has grown rapidly and now exceeds the volume of voice traffic. This situation is forcing service providers to design, build, and add capacity to their networks. Optical networks are providing the necessary infrastructure to meet the need. Due to their high efficiency and straightforward architecture, optical networks can carry large amounts of data over long distances at a reduced cost. This allows network operators to offer more value-added services, such as video conferencing, real-time video on demand, and other high-bandwidth applications, thereby making it easier for the world to communicate at the speed of light.

This chapter begins by giving the motivation for studying optical fibers and looks at how modern optical communication began. It then introduces some basic properties associated with the propagation of light through optical fibers using geometrical-optics description.

1.1 Why optical fiber?

In the mid 1970s, it was recognized that the existing copper technology would be unsuitable for future communication networks. Inherent in copper wire are some problems, including the following:[1]

- Its bandwidth is limited due to physical constraints.
- It is susceptible to radio, electrical, and crosstalk interference, which can garble data transmission.
- Its electromagnetic emissions compromise security.
- Its reliability is reduced in the presence of a harsh environment.

In light of these problems, the telecommunications industry invested heavily in research into optical fibers. Today, optical fiber is becoming the medium of choice for the following reasons:[2–4]

- *High bandwidth*: The quantity of information that can be transmitted by electromagnetic waves increases in proportion to its frequency; therefore, by using light four or five orders of magnitude can be gained in the amount of information transmitted. Light has an information-carrying capacity (or bandwidth) 10,000 times greater than that of the highest radio frequencies. Therefore, optical fiber provides a very high capacity for carrying information; it can be made to carry 10 Tbps. It also has sufficient bandwidth that bit-serial transmission can be used, thereby considerably reducing the size, cost, and complexity of the hardware. The high bandwidth of fiber makes it a perfect candidate for transmitting bandwidth-hungry signals, such as images.
- *High transmission rate*: Fiber optics transmit at speeds much higher than copper wire. Although the current Gbps network is regarded as a milestone compared with the existing Mbps networks, its rate is only a small fraction of the rates possible with fiber optics technology — fiber is capable of transmitting three TV episodes in one second and will be able to transmit the equivalent of an entire 24-volume encyclopedia in one second.
- *Attenuation*: Fiber optics have low attenuation and are therefore capable of transmitting over a long distance (up to 80 km) without the need of repeaters. This low attenuation allows one to extend networks to large campuses or a city and its suburbs.
- *Electromagnetic immunity*: Fiber is a dielectric material. Therefore, it neither radiates nor is affected by electromagnetic interference (EMI), lightning strikes, or surges. The benefits of such immunity include the elimination of ground loops, signal distortion, and crosstalk in hostile environments.
- *Security*: Telecommunication companies need secure, reliable systems to transfer information between buildings and around the world. A fiber-optic network is more secure from malicious interception because the dielectric nature of optical fiber makes it difficult to tap a fiber-optic cable without interrupting communication. Accessing the fiber requires an intervention that is easily detectable by security surveillance. Fiber-optic systems are also easy to monitor. Fiber-optic cables are virtually unaffected by atmosphere conditions. Because the basic fiber is made of glass, it will not corrode or be affected by most chemicals. It can be buried directly in most kinds of soil or exposed to corrosive atmospheres without significant concern. This property of fiber makes it attractive to governments, banks, and others with security concerns.

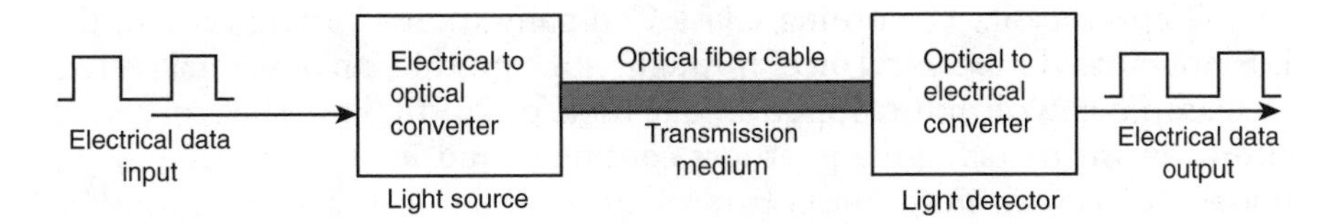

Figure 1.1 A typical optical communication system.

- *Cost:* Glass fibers are made from silica sand, which is more readily available than copper. The cost of optical fibers has fallen considerably over the last few years and will continue to fall. The cost of related components, such as optical transmitters and receivers, is also falling.

These impressive advantages of fiber optics over electrical media have made optical fiber the replacement for copper for fast transmission of enormous amounts of information from one point to another.

However, optical fiber has its drawbacks. First, electrical-to-optical conversion at the sending end and optical-to-electrical conversion at the receiving end is costly. Second, optical fiber suffers the same fate as any wired medium — cable must be buried along the right of way. Third, special installation and repair techniques are required.

A fiber-optic system is similar to a conventional transmission system. As shown in Figure 1.1, a fiber-optic system consists of a transmitter, propagation medium, and receiver. The transmitter accepts and converts input electrical signals in analog or digital form to optical signals and then sends the optical signal by modulating the output of a light source (usually an LED or a laser) by varying its intensity. The optical signal is transmitted over the optical fiber (made of glass or plastic) to a receiver. At the receiver, the optical signal is converted back into an electrical signal by a photodiode.

1.2 A glimpse of history

Guided light has been used for communication purposes for ages. For example, Claude Chappe, the French engineer, invented the "optical telegraph." The advantages of light as a transmission medium were also apparent in the time of Alexander Bell. In fact, Bell invented an optical telephone system known as the Photophone in 1880. However, modern optical communications began with the discovery of the ruby laser in 1960 by T. H. Maimon at Hughes Laboratories in the U.S. This discovery, more than anything else, helped transform our thinking about light as a coherent, continuous electromagnetic radiation. It also led to much research activity in optical communications. In 1966, C. K. Kao and G. A. Hockman, at Standards Telecommunications Laboratories in England, realized that some form of optical waveguide was needed to propagate light and proposed the use of glass fiber. The idea was initially not well received because of the high attenuation of glass fiber. In

1970, Kapron et al., at Corning Glass Company (now Corning Inc.) in the U.S. announced a technical breakthrough — the production of several meters of glass fiber with the required attenuation of 20 dB/km. This concerted effort toward reducing the fiber loss continued and led to obtaining losses under 1 dB/km in pure silica fiber.

As work progressed on reducing fiber loss, laser development continued *pari passu*. The modulation speed with ruby lasers was found to be very low. A faster light source was made available with the discovery of semiconductor lasers in 1962. Gallium arsenide (GaAs) is a semiconductor material that emits light at a wavelength of 870 nm. Semiconductor sources are now available for emitting light at different wavelengths. In addition to lasers, which are quite expensive, low-cost light-emitting diodes (LEDs) were developed.

On the receiver side, the direct detection receiver recovers the information after using a photodetector to convert the optical signal into an electrical one. Early receivers used avalanche photodiodes (APDs). In 1978, D. R. Smith and colleagues at British Telecom Research Laboratories showed that a positive-intrinsic-negative (PIN) photodiode would outperform an equivalent APD receiver. During the mid-1980s, coherent detection techniques, which offer better receiver sensitivity and frequency selectivity than the direct detection schemes, were explored.

The first optical systems were developed using multimode fibers, which propagate many modes, in conjunction with a laser or LED source. A breakthrough came with the development of a single-mode fiber, which propagates a single mode and has lower loss than the multimode fibers. Recently, there is an ongoing discussion on dense wavelength division multiplexing (DWDM), which allows many wavelength channels to be carried by a single fiber.

Optical fiber communication has evolved rapidly and has reached a state of maturity. Optical fiber is available for transmitting practically all types of signals including voice, video, and data, which makes it suitable for local area networks (LANs), metropolitan area networks (MANs), and wide area networks (WANs), where coaxial cable or twisted pair copper wires are being replaced by optical fiber cables. Already the phone companies use fiber-optic facilities, and the TV companies are not lagging behind.

The greatest challenge remaining for fiber optics is economics. While telephone and TV companies can justifiably install fiber links to remote sites, terminal equipment is too expensive to justify installing fiber all the way to homes. Instead, twisted pair or coaxial cables are used to connect optical networks to homes. As the optical revolution continues, we can expect better and better services. The future of optical communication is very bright.

1.3 Optical fibers

An optical fiber is a dielectric cylindrical waveguide operating at optical frequency. Optical frequencies are on the order of 100 THz. Some people question whether the electromagnetic radiation used in fiber should be called

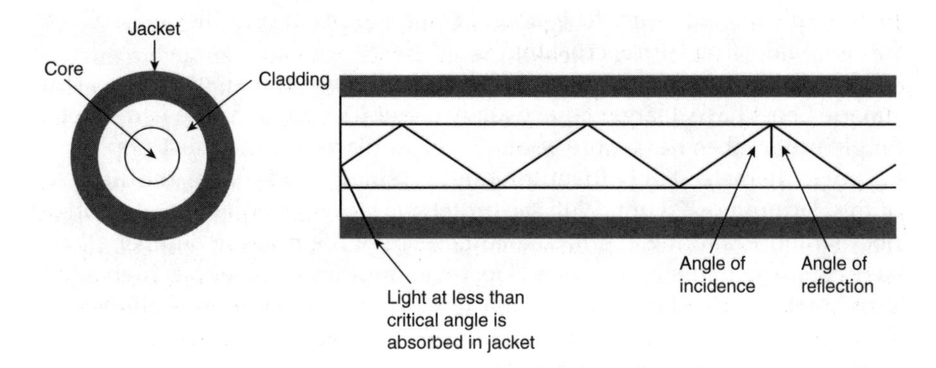

Figure 1.2 Optical fiber.

light since it is outside the visible band of the spectrum. The term is convenient to use because the frequencies involved compel us to consider the quantum nature of the radiation.

As shown in Figure 1.2, an optical fiber consists of three concentric cylindrical sections: the core, cladding, and jacket. The core consists of one or more thin strands made of silica glass (SiO_2) or plastic. The cladding is the glass or plastic coating surrounding the core, which may be step-index or graded-index. In the step-index core, the refractive index is uniform but undergoes an abrupt change at the core-cladding interface, while the graded-index core has a refractive index that varies with the radial distance from the center of the fiber, as illustrated in Figure 1.3. The jacket surrounds one

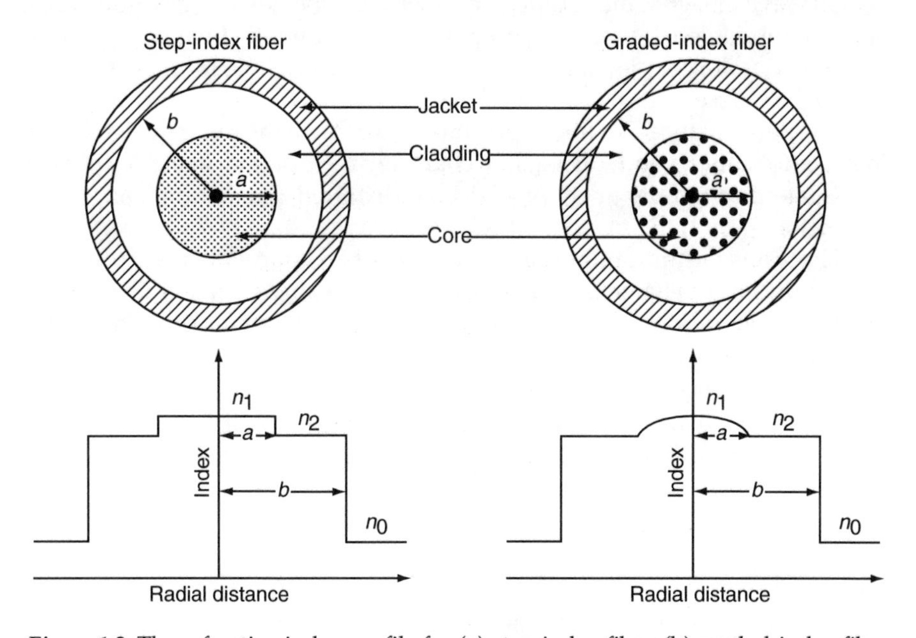

Figure 1.3 The refractive index profile for (a) step-index fiber, (b) graded-index fiber.

or a bundle of clad fibers. The jacket is made of plastic or other materials to protect against moisture, crushing, etc.

There are two major types of fibers: multimode and single-mode. Multimode fibers have a larger core, usually either 62.5 μm or 50 μm in diameter. Single-mode fiber has a core about 8 μm in diameter; the most popular is 8.3 μm in diameter. For both multimode and single-mode fibers, the diameter of the cladding is 125 μm. With the protective jacket, the diameter of a single fiber strand is about 250 μm. The larger core of the multimode fiber allows easy coupling to the light source. However, multimode fiber has higher loss than single-mode fiber and is therefore used only for communications over short distances, such as within a building or on a campus. Single-mode fiber is used for all long-distance communications.

The propagation of light in optical fiber could be examined by Maxwell's equations in cylindrical polar coordinates. However, the mathematics involved with this approach is lengthy and complicated.[5–8] Instead, geometrical-optics will be used to describe the guiding mechanism.

A ray of light entering the core will be internally reflected when incident in the denser medium and the angle of incidence is greater than a critical value. Thus, light travels within the core mostly as a result of total internal reflection at the core-cladding interface. Light rays are reflected back into the core and the process is repeated as light passes down the core. This form of propagation is multimode, referring to the variety of angles that will reflect, as shown in Figure 1.4. It causes signals to spread out in time and limits the rate at which data can be accurately received. By reducing the radius of the core, a single-mode propagation occurs. A mode is a discrete optical wave or signal that propagates down the fiber. In a single-mode fiber, only the fundamental mode can propagate. In multimode fiber, a large number of modes are coupled into the cable, making it suitable for the less costly LED light source.

The performance of a fiber-optic link depends on the numerical aperture (NA), attenuation, and dispersion characteristics of the fiber. As signals propagate through the fiber, they become distorted due to attenuation and dispersion.

The NA is the most important parameter of an optical fiber; a multimode fiber is primarily characterized by its NA. The value of NA is dictated by the refractive indices of the core and cladding. The refractive index n of a medium is defined as

$$\begin{aligned} n &= \frac{\text{speed of light in a vacuum}}{\text{speed of light in the medium}} \\ &= \frac{c}{u} = \frac{\frac{1}{\sqrt{\mu_o \varepsilon_o}}}{\frac{1}{\sqrt{\mu\varepsilon}}} \end{aligned} \tag{1.1}$$

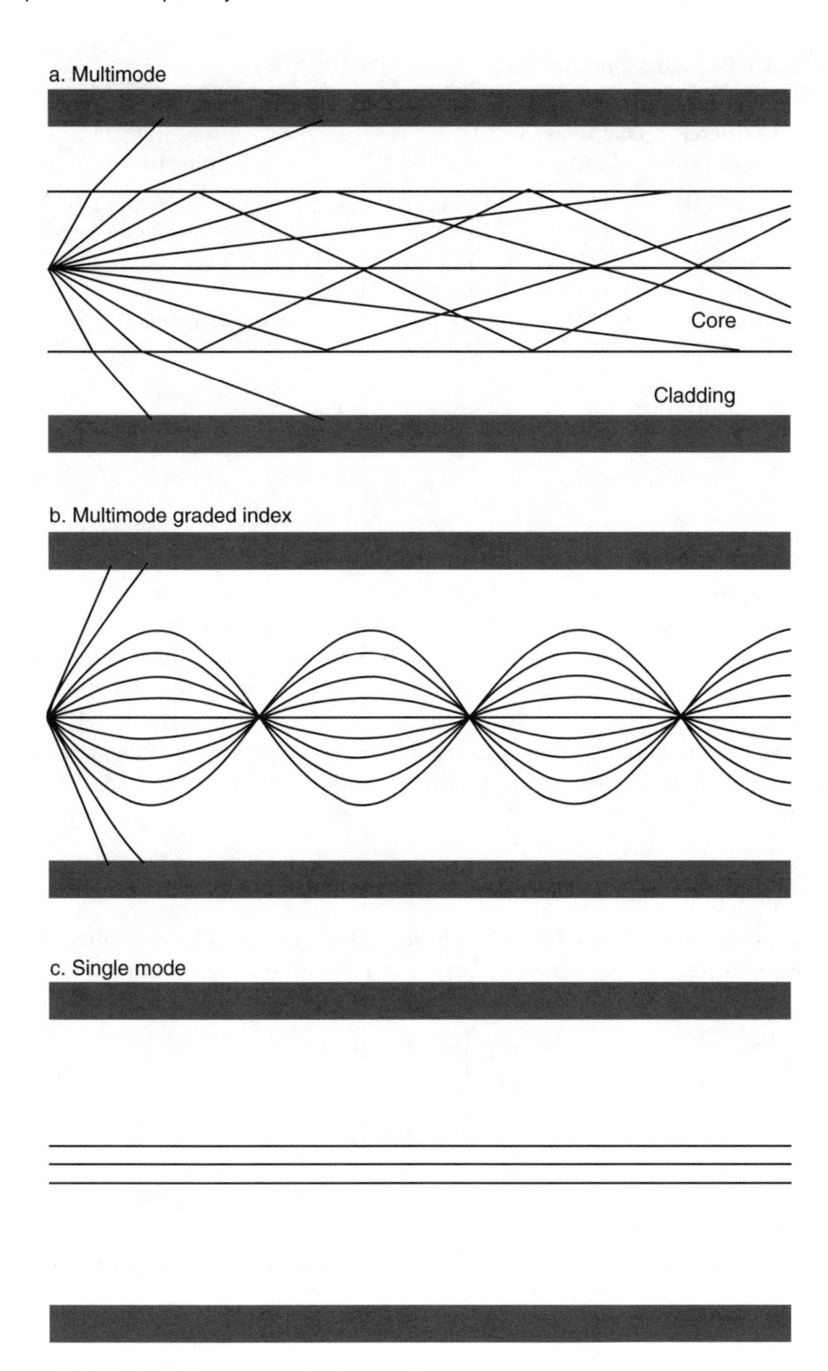

Figure 1.4 Optical fiber transmission modes.

Table 1.1 Standard Optical Fibers

	Core Diameter (μm)	Cladding Diameter (μm)	Δ	Application
I	8–10	125	0.1 to 0.2%	Long distance, high data rate
II	50	125	1 to 2%	Short distance, moderate data rate
III	62.5	125	1 to 2%	LANs
IV	85	125	1 to 2%	LANs
V	100	140	1 to 2%	LANs, short distance

Since $\mu = \mu_o$ in most practical cases, and $\varepsilon = \varepsilon_o \varepsilon_r$

$$n = \sqrt{\frac{\varepsilon}{\varepsilon_o}} = \sqrt{\varepsilon_r} \tag{1.2}$$

indicating that the refractive index is essentially the square root of the dielectric constant. Keep in mind that ε_r can be complex. For common materials, $n = 1$ for air, $n = 1.33$ for water, and $n = 1.5$ for glass at optical frequencies.

Optical fibers for communication purposes are specified and manufactured in five major core/cladding diameters, as shown in Table 1.1, where Δ is the fractional change in refractive index, to be discussed later. An 85/125 fiber means that the fiber has a core diameter of 85 μm and cladding of 125 μm.

1.3.1 Step-index fiber

As mentioned earlier, the refractive index profile of optical fibers can assume various shapes. For step-index fibers, the refractive index assumes

$$n(\rho) = \begin{cases} n_1, & \text{for } \rho < a \\ n_2, & \text{for } \rho > a \end{cases} \tag{1.3}$$

where n_1 and n_2 are the refractive indices of the core and cladding, and a is the radius of the core.

As light propagates from medium 1 to medium 2, Snell's law must be satisfied:

$$n_1 \sin\theta_1 = n_2 \sin\theta_2 \tag{1.4}$$

where θ_1 is the incident angle in medium 1 and θ_2 is the transmission angle in medium 2. The total reflection occurs when $\theta_2 = 90°$, resulting in

$$\theta_1 = \theta_c = \sin^{-1}\frac{n_2}{n_1} \tag{1.5}$$

where θ_c is the *critical angle* for total internal reflection. Note that Equation 1.5 is valid only if $n_1 > n_2$ (to ensure total internal reflection) since the value of $\sin\theta_c$ must be less than or equal to 1. In other words, the cladding must have a lower refractive index than the core.

Another way of looking at the light-guiding capability of a fiber is to measure the *acceptance angle* θ_a, which is the maximum angle over which light rays entering the fiber will be trapped in its core. We know that the maximum angle occurs when θ_c is the critical angle, thereby satisfying the condition for total internal reflection. Thus, for a step-index fiber, the numerical aperture

$$\text{NA} = \sin\theta_a = n_1 \sin\theta_c = \sqrt{n_1^2 - n_2^2} = n_1\sqrt{2\Delta} \tag{1.6}$$

where Δ is the fractional index change at the interface, i.e.,

$$\Delta = \frac{\left(n_1^2 - n_2^2\right)}{2n_1^2} \cong \frac{n_1 - n_2}{n_1} \tag{1.7}$$

n_1 is the refractive index of the core, and n_2 is the refractive index of the cladding, as shown in Figure 1.5. Since most fiber cores are made of silica, $n_1 = 1.48$. For communications applications, most fibers have $\Delta < 0.02$. Typical values of NA range between 0.19 and 0.25. The larger the value of NA, the more optical power the fiber can capture from a source.

The time delay between two rays taking the shortest and the longest paths is a measure of the broadening experienced by an input impulse signal. The shortest path takes place when $\theta_1 = 0$ and is the same as the fiber length L. The longest path takes place when θ_1 satisfies Equation 1.5 and is given by

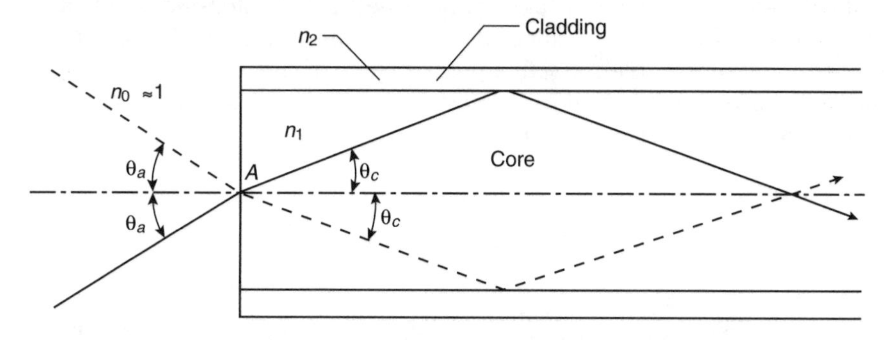

Figure 1.5 Numerical aperture and acceptance angle.

$L/\sin\theta_c$. Dividing the difference in the longest and shortest paths by the velocity in Equation 1.1 gives the time delay as

$$\Delta T = \frac{n_1}{c}\left(\frac{L}{\sin\theta_c} - L\right) = \frac{L}{c}\frac{n_1^2}{n_2}\Delta \tag{1.8}$$

The time delay ΔT can be related to the information-carrying capacity of the fiber, which is measured in terms of the fiber bandwidth or the BL product, where B is the bit rate. We obtain an order-of-magnitude estimate of the fiber bandwidth from the condition $B\Delta T < 1$. Combining this with Equation 1.8 gives a rough estimate of the intrinsic limitation of the step-index fiber

$$BL < \frac{n_2}{n_1^2}\frac{c}{\Delta} \tag{1.9}$$

assuming $a \gg \lambda$, where a is the radius of the core.

Because of the numerous modes a fiber may support, it is called a *multimode step-index* fiber. From electromagnetic field theory, the mode volume V is given by

$$V = \frac{\pi d}{\lambda}\sqrt{n_1^2 - n_2^2} \tag{1.10}$$

where $d = 2a$ is the fiber core diameter and λ is the wavelength of the optical source. From Equation 1.10, the number N of modes propagating in a step-index fiber is finite and can be estimated as

$$N = \frac{1}{2}V^2 \tag{1.11}$$

1.3.2 *Graded-index fibers*

Graded-index fibers have a higher data rate than do step-index fibers. In fact, almost all multimode fibers used today are graded-index fibers. For graded-index fibers, the refractive index profile of the core varies with the radial distance ρ, while that of the cladding is constant. The most common form of refractive index is given by

$$n(\rho) = \begin{cases} n_1\left[1 - \Delta(\rho/a)^\alpha\right], & \text{for } \rho < a \\ n_1(1-\Delta) = n_2, & \text{for } \rho \geq a \end{cases} \tag{1.12}$$

where n_1 is the refractive index at the core axis, n_2 is the refractive index of the cladding, and a is the radius of the core. The parameter α defines the shape of the index profile. For example, $\alpha = 1$ corresponds to a triangular profile, $\alpha = 2$ yields a parabolic profile, and $\alpha = \infty$ simplifies to a step-index profile.

The quantity $\Delta T/L$, where ΔT is the maximum time delay in an optical fiber of length L, is a measure of dispersion (see Section 1.4). It varies with α and has the minimum value when $\alpha = 2(1 - \Delta)$ and

$$\frac{\Delta T}{L} = \frac{n_1 \Delta^2}{8c} \tag{1.13}$$

By applying the criterion $B\Delta T < 1$, we obtain the BL product as

$$BL < \frac{8c}{n_1 \Delta^2} \tag{1.14}$$

The mode volume V for graded-index fibers is defined in the same way as for step-index fibers,

$$V = \frac{2\pi n_1 a \sqrt{2\Delta}}{\lambda} \tag{1.15}$$

The number of modes N can be estimated as

$$N = \frac{V^2}{2}\left(\frac{\alpha}{\alpha + 2}\right) \tag{1.16}$$

The numerical aperture of a graded-index fiber is difficult to determine because the acceptance angle of a ray depends on the radial position of the entry location. However, a local NA is given by

$$\mathrm{NA}(\rho) = \begin{cases} \mathrm{NA}(0)\sqrt{1 - \left(\frac{\rho}{a}\right)^{\alpha}}, & \rho < a \\ 0, & \rho > a \end{cases} \tag{1.17}$$

where NA(0) is the NA at the center of the fiber core. The axial numerical aperture NA(0) is given by

$$\mathrm{NA}(0) = \left[n^2(0) - n_2^2\right]^{1/2} = \left[n_1^2 - n_2^2\right]^{1/2} = n_1\sqrt{2\Delta} \tag{1.18}$$

It is evident from Equation 1.17 that NA for graded-index fiber decreases from the value of NA(0) to zero as the radial distance ρ changes from the fiber axis to the core-cladding interface.

1.4 Fiber loss and dispersion

Besides bandwidth and numerical aperture, two other important properties of optical fibers are loss, or attenuation, and dispersion. Both dispersion and attenuation increase with fiber length, and in conjunction with receiver sensitivity they limit the transmission distance an optical fiber link can cover.

Fiber loss or attenuation plays a major role in the design of an optical communication system since it sets a limit on the maximum distance between a transmitter and a receiver or between signal repeaters. As light propagates down the optical fiber, it loses power over distance. Power attenuation (or fiber loss) in an optical fiber is governed by

$$\frac{dP}{dz} = -\alpha P \tag{1.19}$$

where α is the attenuation and P is the optical power. In Equation 1.19, it is assumed that a wave propagates along z. By solving Equation 1.19, the powers $P(0)$ and $P(L)$ at the entry and exit of the fiber, respectively, are related as

$$P(L) = P(0)e^{-\alpha L} \tag{1.20}$$

It is customary to express attenuation α in dB/km and length L of the fiber in km. In this case, Equation 1.20 becomes

$$\alpha = \frac{10}{L}\log_{10}\frac{P(0)}{P(L)} \tag{1.21}$$

Thus, the power of the light reduces by α dB/km as it propagates through the fiber. Equation 1.21 may be written as

$$P(L) = P(0) * 10^{-\alpha L/10} \tag{1.22}$$

For $L = 100$ km,

$$\frac{P(L)}{P(0)} \cong \begin{cases} 10^{-100}, & \text{for coaxial cable} \\ 10^{-2}, & \text{for fiber} \end{cases} \tag{1.23}$$

indicating that more power is lost in the coaxial cable than in fiber.

Fiber losses are due to two major factors: material absorption and Rayleigh scattering effects, both of which are wavelength dependent. Scattering is pronounced at short wavelengths, while absorption dominates at long wavelengths. Material absorption includes absorption by the fiber material

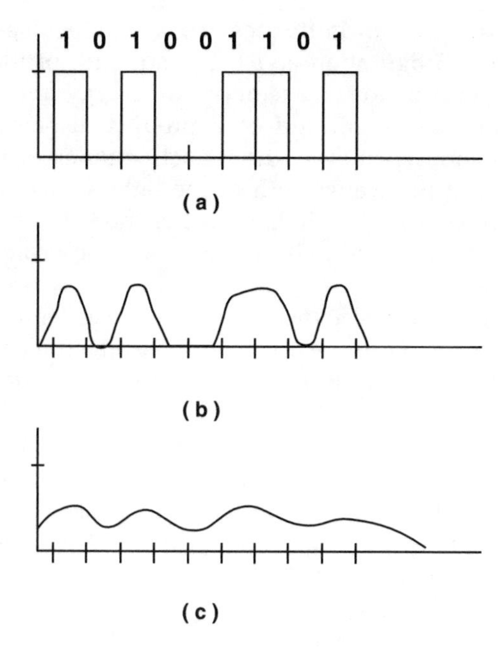

Figure 1.6 Dispersion of digital optical signal: (a) no dispersion, (b) acceptable dispersion, (c) extensive dispersion causing an unacceptable signal.

as well as by the impurities in the material. Rayleigh scattering arises from fluctuations in the density of the material at the microscopic level. For standard glass fiber, there is a lower limit of about 0.15 dB on the attenuation per kilometer due to Rayleigh scattering. Thus, the signal needs to be amplified for long-distance transmission.

Dispersion is the phenomenon that limits the data rate of any optical fiber. It is essentially the spreading of light pulses as they propagate down a fiber. As the pulses, represented by zeros, spread, they overlap epochs that are represented by ones (1), as in Figure 1.6. If dispersion is beyond a certain limit, it may confuse the receiver. The dispersive effects in single-mode fibers are much smaller than those in multimode fibers.

A typical effect of dispersion is illustrated in Figure 1.6. The original digital optical pulses are discrete; ones and zeros are easily identified, as in Figure 1.6(a). After some distance, dispersion occurs, but the pulses can still be decoded, as in Figure 1.6(b). Further propagation of the signal makes it become totally distorted and unacceptable because of extensive dispersion, as in Figure 1.6(c).

There are two types of dispersion: intermodal and intramodal. Intermodal dispersion (also known as multimode) is the propagation of rays along different paths through an optical fiber and their arrival at the end of the fiber at different times. Intramodal dispersion is caused by three factors: material dispersion, waveguide dispersion, and cross-product dispersion. Material dispersion occurs because of the change in the refractive index of a fiber material with the optical frequency. Waveguide dispersion is caused by the waveguide configuration and the bandwidth of the signal. Cross-product dispersion is due to leakage of optical energy from one material to others. Since both wavelength and cross-product dispersions are usually negligible, dispersion depends on the refractive index of the material the fiber is made of and the wavelength of the light source. Consequently, a medium is said to be dispersive if its refractive index is a function of wavelength (or frequency). Dispersion is measured in nanoseconds per kilometer (ns/km).

The bandwidth of graded-index multimode fibers tends to be limited by the modal dispersion. Thus, many high-data-rate transmission systems switch from graded-index multimode fibers to single-mode fibers.

Summary

Since the invention of optical fiber in the early 1960s, its use and demand have grown tremendously. The uses of optical fiber are numerous. The most common are found in telecommunications, medicine, military, and industry. As the demand for higher bandwidth and transmission rate increases, optical fiber will continue to play a vital role in telecommunications.

An optical fiber is a cylindrical waveguide consisting of a glass or plastic core surrounded by a lower-index cladding, and the whole structure is protected by a jacket. The optical characteristics of the fiber are dictated by its refractive index profile.

The major properties of the fiber include bandwidth, numerical aperture, attenuation (NA), and dispersion. Bandwidth is the information-carrying capacity of the fiber. The numerical aperture of an optical fiber, which cannot be greater than 1.0, is the sine of the acceptance angle. Fiber loss or attenuation is the reduction in the power of the optical signal. Dispersion is the spreading of light rays which can cause pulses representing different information symbols to overlap and become inseparable at the receiver. The properties originate from the geometric and physical nature of the fiber and have a profound limiting effect on the information-carrying capacity of the fiber.

References

1. J. Enck and M. Beckman, *LAN to WAN Interconnection,* McGraw-Hill, New York, 1995, 127–153.
2. M. N. O. Sadiku, *Metropolitan and Wide Area Networks,* Prentice-Hall, Upper Saddle River, NJ, in press.
3. S. P. Joshi, High-performance networks: A focus on the fiber distributed data interface (FDDI) standard, *IEEE Micro,* vol. 6, no. 3, June 1986, 8–14.
4. R. Jain, *FDDI Handbook: High-Speed Networking Using Fiber and Other Media,* Addison-Wesley, Reading, MA, 1994.
5. M. J. N. Sibley, *Optical Communications,* McGraw-Hill, New York, 1990, 8–39.

6. J. E. Midwinter and Y. L. Guo, *Optoelectronics and Lightwave Technology,* John Wiley & Sons, Chichester, 1992, 179–187.
7. R. Papannareddy, *Introduction to Lightwave Communication Systems,* Artech House, Boston, 1997, 22–32.
8. G. P. Agrawal, *Fiber-Optic Communication Systems,* 2nd ed., John Wiley & Sons, New York, 1997, 30–39.

Problems

1.1 What advantages does optical fiber have over copper wire as a transmission medium? Describe an application in which one would prefer using optical fiber.

1.2 A standard glass fiber has a core index of 1.45 and an acceptance angle of 42°. Find the cladding index.

1.3 Assume a step-index fiber with $n_1 = 1.456$ and $n_2 = 1.45$ communicating data at a bit rate of 10 Mbps. Find the maximum length of the fiber.

1.4 A step-index fiber has a core radius of 16 μm with $n_1 = 1.56$ and $\Delta = 2 \times 10^{-3}$. At $\lambda = 1.55$ μm, calculate the NA, acceptance angle, and the number of propagating modes.

1.5 For a step-index multimode fiber, $n_1 = 1.458$, $\Delta = 1.2\%$, $\lambda = 1.52$ μm, and $a = 25$ μm. Calculate NA and V.

1.6 Show that the formula

$$\frac{\sin\theta_a}{\sin(90° - \theta_c)} = \frac{n_2}{n_1}$$

is the same as

$$\sin\theta_a = n_2\sqrt{1 - \sin^2\theta_c}$$

when medium 1 is air.

1.7 Show that for step-index fibers,

$$d = 2a \cong \frac{0.22\lambda}{\text{NA}}$$

1.8 For $0 < \rho < a$, plot $n(\rho)$ for a graded-index fiber with $\alpha = 1,2,4,8$, and ∞ (step index). Let $a = 25$ μm, $n_1 = 1.46$, and $\Delta = 0.02$.

1.9 Consider a graded-index fiber operating at 1300 nm. If $a = 20$ μm, $\alpha = 2$, $n_1 = 1.46$, and $n_2 = 1.45$, determine the number of modes in the fiber. How does this compare with the number of modes in a step-index fiber with identical characteristics?

1.10 A lightwave system is characterized by an attenuation of 2.4 dB/km. If 0.6 mW is launched into the fiber, how much power can be received at 5 km?

1.11 An optical signal loses 77% of its power after going through 600 m of a certain optical fiber. Calculate the fiber loss in dB/km.

chapter two

Optical transmitters and receivers

People avoid change until the pain of remaining the same is greater than the pain of changing.

— Unknown

The essential components of an optical fiber communication system are a transmitter, a transmitting medium (optical fiber), and a receiver. The transmitter and receiver may be either single components or in an array form. In fiber optics, the role of the transmitter is threefold: convert input electrical signal into a diode drive current, convert the current into an optical signal, and launch it into the optical fiber acting as a transmission medium. The role of the receiver is to convert the optical signal back to the original electrical signal, thereby recovering the data sent by the transmitter. This chapter discusses optical transmitters and receivers and introduces their various components.

2.1 *Optical sources*

The purpose of an optical source is to convert electrical signals to light waves. Optical sources designed for communication purposes must meet the following criteria:[1,2]

- Have a small emitting area comparable to the size of the fiber core
- Emit light within the acceptance cone to ease coupling to the fiber
- Emit the most advantageous wavelength for propagation
- Be intense
- Have rapid response time
- Be reliable and economical
- Be easily modulated and capable of high speed modulation
- Be able to operate continuously at room temperature for a long period

Table 2.1 Typical Semiconductors Used in Fiber-Optic Devices[3]

Symbol	Name
Al	Aluminum
$Al_xGa_{1-x}As$	Aluminum-gallium-arsenide
As_2	Arsenic
AsH_3	Arsine
Ga	Gallium
GaAs	Gallium arsenide
GaN	Gallium nitride
Ge	Germanium
In	Indium
InAs	Indium arsenide
InGaAlAs	Indium-gallium-aluminum-arsenide
$In_{1-x}Ga_xAs$	Indium-gallium-arsenide
$In_{1-x}Ga_xAs_yP_{1-y}$	Indium-gallium-arsenide-phosphide
InGaN	Indium-gallium-nitride
InP	Indium phosphide
$LiNbO_3$	Lithium niobate
P	Phosphorous
PH_3	Phosphine
Si	Silicon
SiGe	Silicon germanium
SiO_2	Silicon dioxide; silica

Although several devices can convert electrical signals to light waves, only two sources satisfy these requirements and are suitable for fiber-optic communication systems: (1) the light-emitting diode (LED) and (2) the injection laser diode (ILD). The two emitters have the required low voltage (or low power requirement), compact size, low cost, high efficiency, reliability, and desired wavelength range for an optical fiber. They have much in common. They both operate in forward-biased mode. They are also cousins in that they are made of the same semiconductors such as gallium-aluminum-arsenide-phosphide (GaAlAsP) and indium-gallium-arsenide-phosphide (InGaAsP). Table 2.1 lists some of the semiconductors used in fiber-optic devices.

A basic difference between an LED and an ILD is that the output from an LED is incoherent, whereas that from an ILD is coherent. Thus, the coherent output from an ILD can be coupled into either single-mode or multimode fibers, whereas the incoherent output from an LED can only be coupled to a multimode fiber.

Light sources can be classified as short-wavelength and long-wavelength sources producing light from 500 to 1000 nm and from 1200 to 1600 nm, respectively.

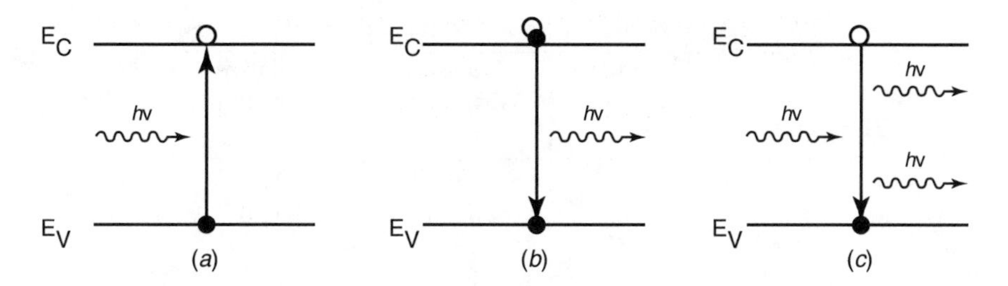

Figure 2.1 Three processes of light generation: (a) spontaneous emission, (b) stimulated emission, (c) absorption.

2.1.1 Basic concepts

As illustrated in Figure 2.1, the basic operation of light sources involves three processes: (1) spontaneous emission, (2) stimulated emission, and (3) absorption. In Figure 2.1, the gap between the valence bond (ground state) and the conduction band (excited state) is known as the *bandgap*. The gap represents the energy (or energy gap) E_g required by an electron to jump from valence bond to conduction band. Light is emitted during spontaneous emission and stimulated emission.

Spontaneous emission occurs when an excited electron wants to return to the valence bond. To ensure that energy is conserved, the falling of the electron to a lower energy level causes photons to be emitted in random directions, thereby producing incoherent light. LEDs emit incoherent light through the process of spontaneous emission.

Stimulated emission takes place when the electron is in the excited state with excess energy. When stimulated by an external photon (incident light), the electron drops to the valence bond while releasing a photon with the same amount of energy as the external photon. This results in coherent light. Stimulated emission can occur only when there are more electrons in the conduction band than in the valence band. In other words, the conduction band must have a large number of electrons, while the valence band must contain a large number of holes. All semiconductor lasers emit light through the process of stimulated emission.

Absorption takes place only when the photon energy (of the incident light) $\hbar f = E_c - E_v$, where $f = c/\lambda$ is the frequency, E_c is the energy of the conduction band, and E_v is the energy of the valence band. In this case, the photon is absorbed by the electron, which ends up in the conduction state. The incident light is absorbed and there is no emission.

The semiconductor used for an optical source must have a direct bandgap. In an indirect bandgap material, the energy involved in electron-hole recombination is released as heat, whereas in a direct bandgap semiconductor material, either heat or light is produced. None of the single-element semiconductors has a direct bandgap. However, compounds of elements III

Table 2.2 Properties of Typical Semiconductor Materials

Material	Bandgap Energy (eV)	Emission Wavelength (nm)	Relative Permittivity ε_r
Si	1.11	1100	11.8
Ge	0.67	1850	16.0
GaAs	1.43	870	13.2
GaP	2.26	2260	11.1
GaSb	0.7	1720	15.7
InAs	0.36	3440	14.6
InP	1.28	920	12.4

and IV of the periodic table are direct-gap materials. These so-called III-IV semiconductors are made of a group III element (such as Al, Ga, or In) and a group IV element (such as P, As, or Sb). Different binary, ternary, and quaternary combinations of the elements (such as GaAs, InAs, AlGaAs and InGaAsP) are direct-gap material.

When an electron drops from the conduction band to the valence band, the energy lost is the same as the photon of light. Applying the basic quantum-mechanical relationship between the bandgap energy E_g and frequency v gives

$$E_g = \hbar f = \frac{\hbar c}{\lambda} \tag{2.1}$$

where $\hbar = 6.62 \times 10^{-34}$ J.s is Planck's constant, $c = 3 \times 10^8$ m/s is the speed of light, and λ is the emission in wavelength. If the bandgap energy is expressed as electron volts (eV), then Equation 2.1 becomes

$$\lambda(\text{nm}) = \frac{\hbar c}{eE_g} = \frac{1244}{E_g(\text{eV})} \tag{2.2}$$

where $e = 1.6 \times 10^{-19}$ C is the electronic charge. The bandgap energy and emission wavelength of typical semiconductor materials are listed in Table 2.2.

A ternary material is often represented in the form of $Ga_xAl_{1-x}As$, where x is the fraction of gallium in the Ga-Al combination. The energy gap in eV is given by the empirical equation

$$E_g = 1.424 + 1.266x + 0.266x^2, \qquad 0 \le x \le 0.37 \tag{2.3}$$

A quaternary material is represented typically as $In_{1-x}Ga_xAs_yP_{1-y}$, where x and $(1-x)$ denote the proportions of group III elements, while y and $(1-y)$

represent the proportions of group IV elements. The bandgap energy in eV for this semiconductor is

$$E_g = 1.34 - 0.72y + 0.12y^2, \qquad 0 \le x \le 0.47 \tag{2.4}$$

where $y = 2x$. For example, if x = 0.26, the semiconductor $In_{0.74}Ga_{0.26}As_{0.56}P_{0.44}$ has the bandgap energy of 0.96 eV from Equation 2.4. Using Equation 2.2, this alloy emits light at λ =1.3 μm.

2.1.2 Light-emitting diodes (LEDs)

The semiconductor LED is best suited for optical systems requiring bit rates of less than 50 Mbps. It has a long life span (10^6 h), operational stability, low temperature operation, and is low cost.

The LED is an incoherent source of light that emits light by spontaneous emission. It propagates many modes, and is thus known as a multimode source, which makes it suitable for multimode fibers. Typical symbols for LEDs are shown in Figure 2.2. As shown in Figure 2.3, there are two types of semiconductor LEDs in common use: surface emitters and edge emitters.

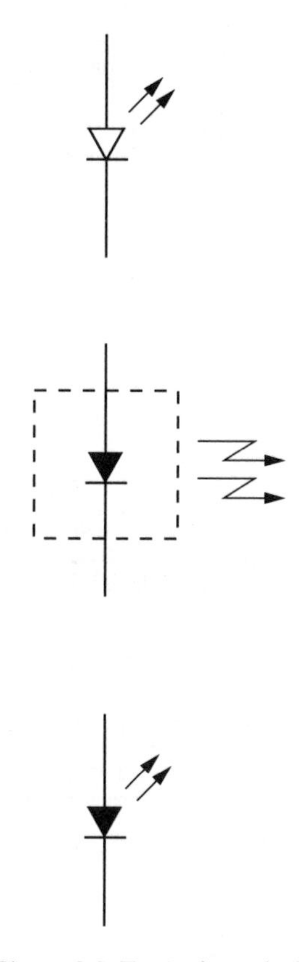

Figure 2.2 Typical symbols for LEDs.

The surface-emitting LED (also called a front or Burrus emitter), developed by C. A. Burrus at Bell Labs, consists of a well or dip etched through the GaAs substrate. The well reduces the size of the light-emitting area. The fiber is epoxied into the well to transmit the light. Light from the light-emitting region can be collected from either side of the device. The output is taken from the light passing through one of the large-area surfaces. Because of the circular symmetry of the emitting region, the emitted beam pattern is also symmetric. Efficient coupling is needed between the LED and the fiber because the LED emits light in many directions.

The edge-emitting LED, developed at RCA, emits light from the edges. It is a heterojunction device in that it uses three or more layers. The upper and lower layers with a lower index of refraction sandwiches the center layer. The change in the refractive index across the heterojunction serves to constrain some of the emitted light to the active region. How the emitted light is coupled to the fiber determines how much power can be launched. In order to make the

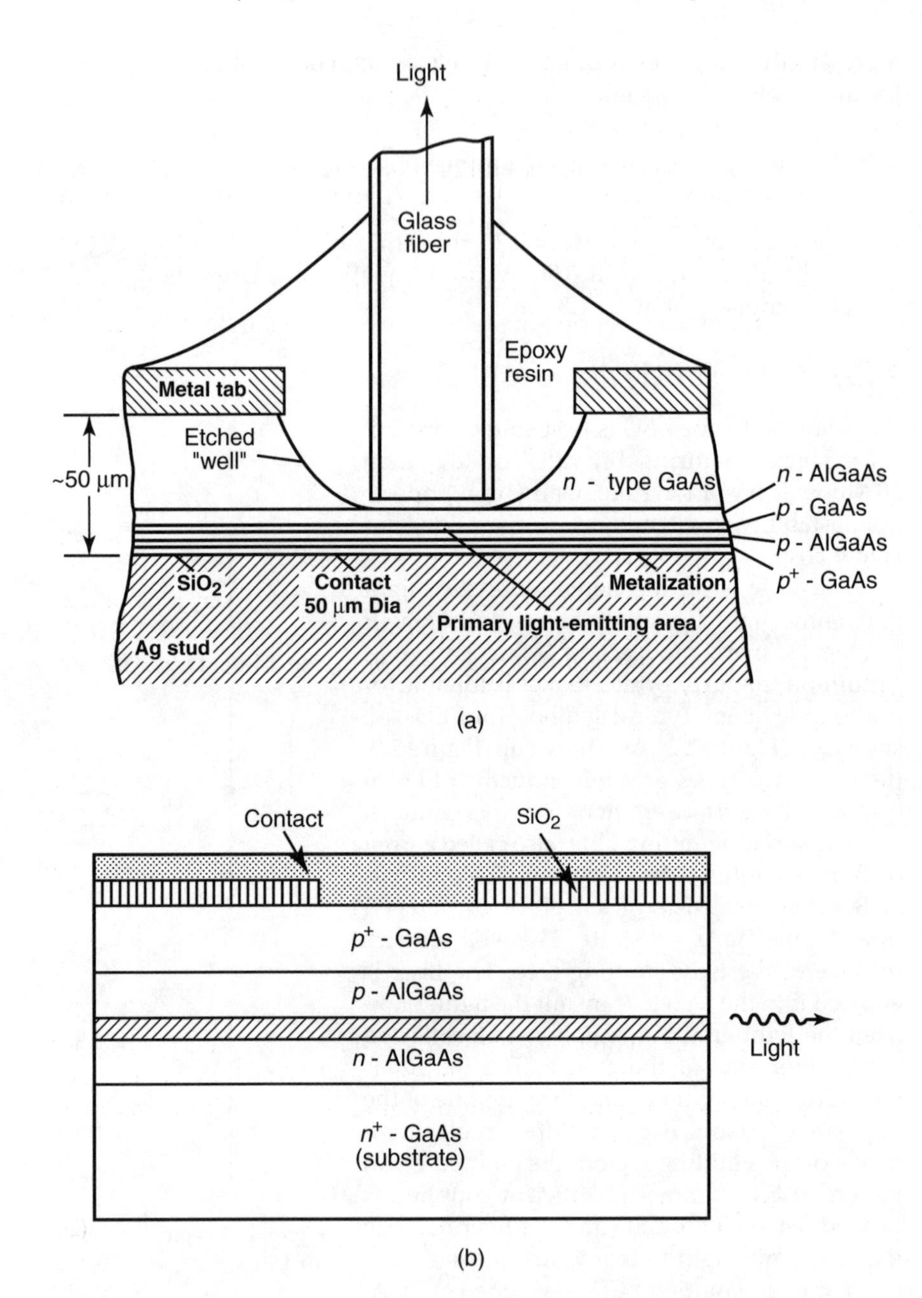

Figure 2.3 Two common LED configurations: (a) surface-emitting, (b) edge-emitting.

beam directional, the light is taken from the edge of the LED, which leads to an effective power launch. Because it is possible to confine the radiating portion to a spot, the edge-emitting LED provides an excellent match to small-diameter fibers.

The typical size of the edge-emitting LED is 10 × 300 μm. For the surface-emitting LED, the diameter of the active region is comparable to that of the optical fiber to which it is coupled. The edge-emitting LEDs are superior to surface-emitting LEDs in terms of coupled power and maximum modulation frequency. Therefore, surface emitters are commonly used for short-haul, edge emitters for medium-haul, and lasers for long-haul routes.

An important characteristic of an LED is the power generated by it because the higher the power, the longer the distance can be between repeaters. The output power is approximately linearly related to the drive current. The optic output power $p(\omega)$ is given by

$$p(\omega) = \frac{p(0)}{\sqrt{1+\omega^2\tau^2}} \tag{2.5}$$

where $p(0)$ is the output power at zero modulation frequency and τ is the time constant of the LED and drive circuit. With a carefully designed drive circuit, the time constant of the LED will be dominant. From Equation 2.5, note that at frequency $\omega = 1/\tau$, the AC power is reduced by 0.707. For this reason, $1/\tau$ is called the *modulation bandwidth* of the LED. Thus, the 3-dB bandwidth in hertz is

$$f_{3\text{dB}} = \frac{1}{2\pi\tau} \tag{2.6}$$

When there are both radiative and nonradiative recombinations, τ is given by

$$\frac{1}{\tau} = \frac{1}{\tau_r} + \frac{1}{\tau_{nr}} \tag{2.7}$$

where τ_r and τ_{nr} are radiative and nonradiative lifetimes, respectively. These constants also give us the *internal quantum efficiency,* which is the quality of the conversion from electrical current to light. The internal quantum efficiency η is given by

$$\eta = \frac{\tau_{nr}}{\tau_{nr} + \tau_r} \tag{2.8}$$

In order to produce a fast LED, both τ_r and τ_{nr} should be made low and where $\tau \gg \tau_{nr}$ so that efficiency is high.

2.1.3 *Laser diodes*

The term *laser* is an acronym for **l**ight **a**mplification by **s**timulated **e**mission of **r**adiation. Lasing is the state at which light amplification is made possible

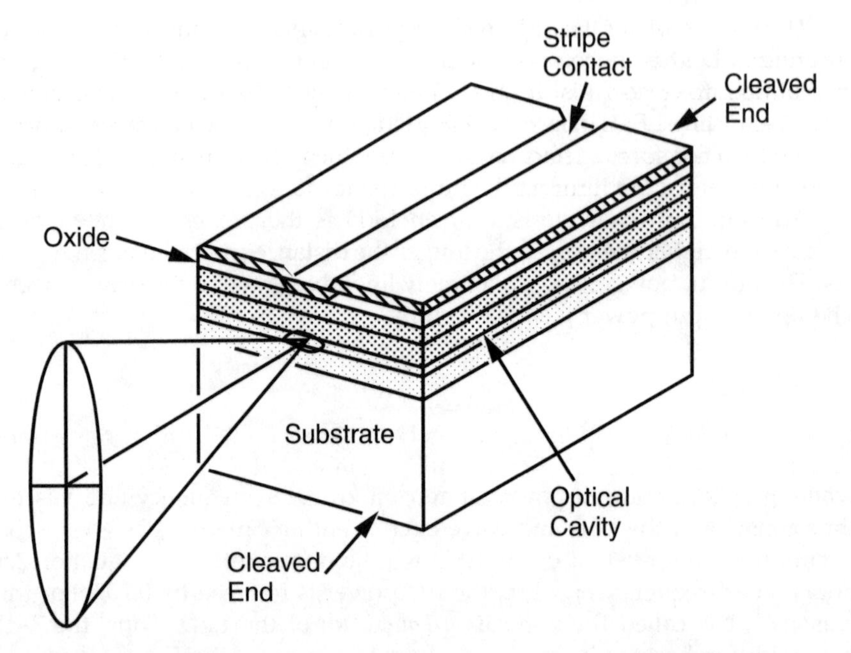

Figure 2.4 A typical injection laser diode (ILD).

in a laser diode. The injection laser diodes (ILDs) and LEDs are different in many respects. ILDs produce 10 dB more power than LEDs. Unlike LEDs, which emit light by spontaneous emission, ILDs produce light by stimulated emission and are mainly edge emitters. While the LEDs produce incoherent light, the ILDs produce coherent or monochromatic light. By coherence, we mean that all emitted light is of the same wavelength.

A typical laser diode is illustrated in Figure 2.4. An ILD can use the same semiconductor material as an LED, but the structure is similar to an edge-emitting LED. Because the active region is sandwiched between layers with lower refractive index, emitted light can exit only the end faces of the ILD.

The laser process is achieved through optical resonance. The active region forms a resonant cavity that resembles the Fabry-Perot cavity used for gas lasers. There are two types of laser structures: gain-guided and index-guided lasers.

In a *gain-guided* structure, the current injection is limited over a narrow stripe. The optical gain is also maximum at the center of the stripe. The index of refraction of the material is so high that the ends make effective mirrors. This parallel mirror structure is known as a *Fabry-Perot interferometer.* Light is confined to the stripe region and this is aided by gain. Consequently, such lasers are called *gain-guided semiconductor lasers.*

The *index-guided* structure incorporates a deliberate change in the material in the lateral direction. That is, a discontinuous change in the refractive index (step index) is built in the structure in the lateral direction. The change in the index of refraction of the material confines the light to a narrower region. Because of the large built-in index step, the spatial distribution of the emitted light is inherently stable.

Some characteristics of ILDs are noteworthy. ILDs are much faster than LEDs because they are based on stimulated emission, which has a shorter lifetime than the spontaneous emission on which LEDs are based. The faster stimulated emission process ensures that a laser diode gives a faster response.

The peak emission wavelength is given by Equation 2.2, namely,

$$\lambda(\mu\text{m}) = \frac{1.244}{E_g(\text{eV})} \tag{2.9}$$

where E_g is the bandgap energy in eV. For example, the bandgap energy of the alloy $Ga_{0.93}Al_{0.07}As$ is 1.51 eV so that it emits at $\lambda = 0.82$ µm.

In laser diodes, light emission takes place in a rectangular cavity. The propagating wave can be expressed in terms of the electric field phasor as

$$E(x,t) = E_o \exp(-\alpha x/2)\exp[j(\omega t - \beta x)] \tag{2.10}$$

where α is the attenuation constant and β is the phase constant in the active region. If we assume that one mirror is located at $x = 0$ and the other at $x = L$ and that the mirrors have reflection coefficients R_1 and R_2, the field at $x = 0$ is

$$E(0,t) = E_o \exp(j\omega t) \tag{2.11}$$

If the wave in Equation 2.10 is launched at $x = 0$, it undergoes a round trip of distance $2L$ and is amplified by stimulated emission. Upon returning to the starting point, the wave becomes

$$E(0,t) = \sqrt{R_1 R_2} E_o \exp[(g-\alpha)L]\exp[j(\omega t - 2\beta L)] \tag{2.12}$$

where g is the power gain per unit length. For steady-state oscillation in the cavity, the phase of the returned wave must equal that of the original wave. Thus, from Equations 2.11 and 2.12,

$$\exp(-j2\beta L) = 1 \rightarrow 2\beta L = 2\pi N \tag{2.13}$$

where N is an integer. Substituting $\beta = 2\pi n/\lambda$, where n is the refractive index, Equation 2.13 becomes

$$\lambda = \frac{2nL}{N} \tag{2.14}$$

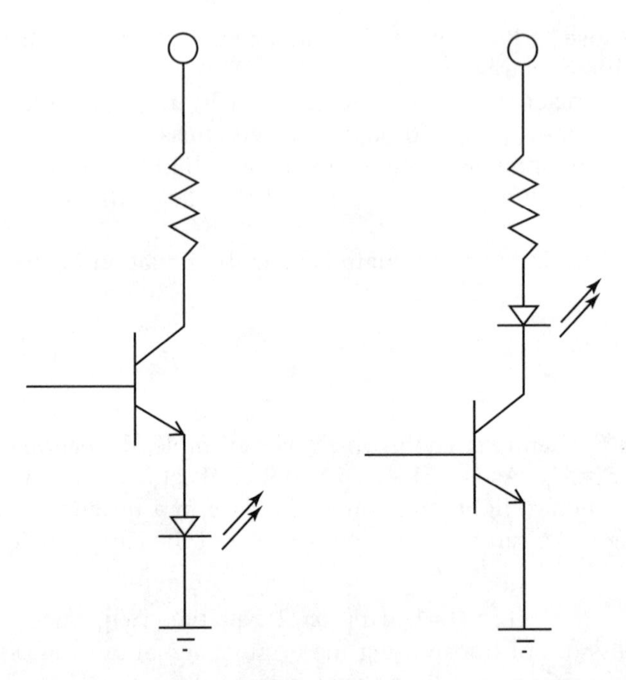

Figure 2.5 A driving circuit using bipolar transitor.

This implies that the laser will amplify only wavelengths that satisfy Equation 2.14. In other words, the cavity length must be an integral number of possible half wavelengths. Note that the result in Equation 2.14 is based on the assumption that a uniform plane wave propagates in the resonant cavity.

2.2 Optical transmitters

Although an optical source is a basic component of an optical transmitter, it is not the only component of the transmitter. Other components include a driving circuit, modulator, source coupler, and multiplexer.

The driving circuit is responsible for supplying current (or electrical power) to the optical source. A typical driving circuit for a laser transmitter is shown in Figure 2.5.

The role of the coupler is to provide maximum transfer of optical energy from the optical source to the fiber and to launch the optical signal into the fiber for transmission. Fiber ends must be precisely prepared, and alignment must be accurate for efficient coupling of the light source to the fiber. Since coupling can be inefficient, the coupling efficiency is defined as

$$\eta = \frac{P_f}{P_s} \tag{2.15}$$

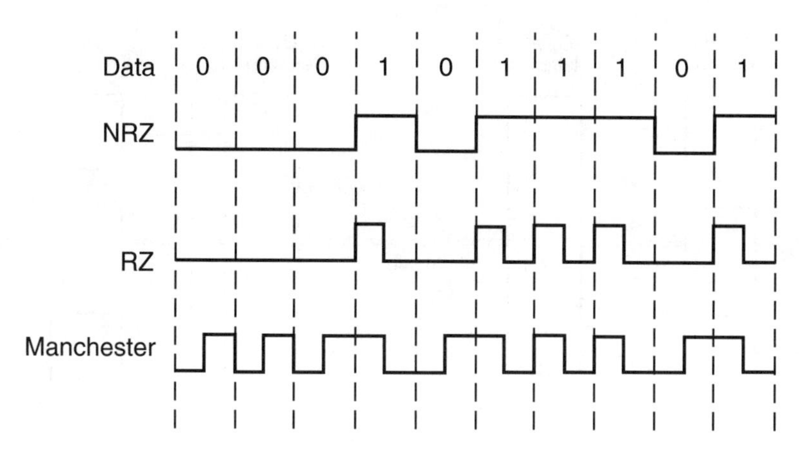

Figure 2.6 Three popular coding schemes.

where P_s is the power produced by the light source and P_f is the power in the fiber. The efficiency is determined by the radiation pattern of the light source and the numerical aperture of the fiber.

The modulator converts electrical data into an optical signal. Modulation is the process of not transmitting the original signal but transmitting it as amplitude, frequency, or phase changes on another signal called the *carrier*, producing amplitude modulation (AM), frequency modulation (FM), or phase modulation (PM), respectively. Carrier modulation is rarely used with fiber optics because it demands that the optical source be treated in a continuous linear manner.

Pulse code modulation (PCM) is another modulation technique used in fiber optics. It is the scheme that takes the analog signal and transforms it to a digital form. It accomplishes that transformation by sampling the signal at regular intervals, quantizing the sample (or rounding it off to the nearest level), and assigning its binary equivalent. The telephony industry used PCM to multiplex voice signals.

Whether analog carrier modulation or PCM is used, the optical transmitter transmits the signal by varying the amplitude of the output power. There are various signal coding schemes used in optical communication systems; three common schemes are non-return-to-zero (NRZ), return-to-zero (RZ), and Manchester, illustrated in Figure 2.6. The choice of input waveform and modulation scheme is dictated by the performance of various optical components. For example, to modulate the semiconductor laser, a current greater than the threshold current is employed and varied according to the modulating signal. The modulation can be analog or digital, pulse or continuous-wave.

The role of the multiplexer is to transmit two or more channels simultaneously over the same fiber. The two popular multiplexing schemes used in fiber optics are time-division multiplexing (TDM) and wavelength-division multiplexing (WDM). TDM is the process of combining many channels

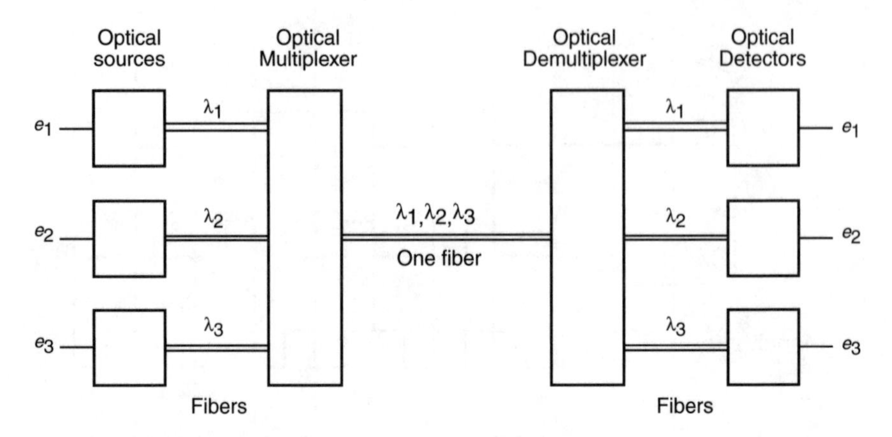

Figure 2.7 Wavelength-division multiplexing (WDM).

of information within a single transmission channel by assigning each channel a different time slot. TDM is rarely used in fiber optics; it is more common in telephone (voice) systems. WDM is a scheme that combines several channels into a single fiber by assigning each channel a different wavelength. WDM is specific to fiber optics. As shown in Figure 2.7, several optical sources are transmitted at different wavelengths. The various wavelengths are injected into a single optical fiber. The light is filtered into separate wavelengths and converted to their respective electrical signals by the receiver.

2.3 Optical detectors

Optical detection is the function of the optical receiver that is unique to optical communication systems. An optical detector (or photodetector) does the opposite of what the optical source does; it converts the normally weak optical input signal into a current output. Optical detectors are usually photodiodes that are photoelectric devices. The range of wavelength detected includes ultraviolet, visual, and infrared, i.e., wavelengths from 0.005 to 4000 μm.

Semiconductor photodiodes are small, light, and fast and can operate with just a few bias volts. Semiconductor materials commonly used for making photodiodes are Si, Ge, GaAs, InAs, and InGaAs. Indirect bandgap materials, such as Si and Ge, are preferred over direct bandgap materials.

Just as optical sources must satisfy certain requirements, photodiodes must have the following characteristics because of their crucial role;[1,4] they must

- Be highly sensitive (or responsive) to weak light signals
- Respond uniformly to all wavelengths
- Be fast enough to transform the light to electrons

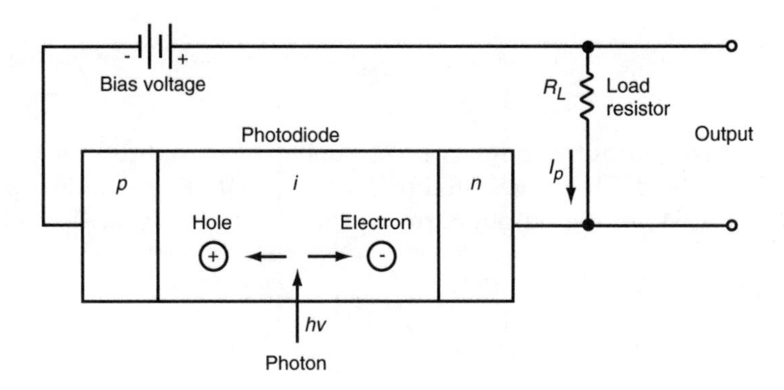

Figure 2.8 A typical PIN photodiode.

- Have a small dimension and be compatible with the fiber
- Have sufficient bandwidth to accommodate the information rate
- Be unaffected by the environment
- Add no noise to the signal
- Be inexpensive

There are two types of photodiodes: positive-intrinsic-negative (PIN) photodiode and avalanche photodiode (APD). PIN photodetectors are used as receivers for LEDs for lower modulation frequencies and short transmission distances. APDs are used for high modulation frequencies and long transmission distances.

2.3.1 PIN photodiode

The PIN photodiode is the most common photodetector. It is made of a semiconductor *pn* junction with reverse bias. The reverse bias voltage is about 7 to 10 V. As shown in Figure 2.8, it has a lightly *n*-doped intrinsic layer sandwiched between the *p*- and *n*-type regions. With the addition of the intrinsic layer, the active region of the diode is increased considerably. Light enters the diode through a tiny "window" about the same size as the fiber core. Light falls on the *p* layer and is absorbed by an electron. The detector does not convert the light to current below a certain wavelength known as the *cutoff wavelength* λ_c, which is related to the bandgap energy E_g in eV as

$$\lambda_c = \frac{hc}{E_g} = \frac{1.24}{E_g} \ (\mu\text{m}) \tag{2.16}$$

An incident optical power P_o with frequency f produces $p_o/\hbar f$ photons per second. Since *quantum efficiency* η is the ratio of the average number of output electrons to the number of incident photons, the average number of electrons per second is $\eta P_o/\hbar f$. The output photocurrent is therefore

$$I_o = \frac{e\eta P_o}{\hbar f} \tag{2.17}$$

A photodiode's output current is proportional to its output current. The sensitivity of the diode is measured by its *responsivity* R. We define responsivity R (in A/W) as the output current produced per unit incident power,

$$R = \frac{I_o}{P_o} = \frac{e\eta}{\hbar f} \tag{2.18a}$$

or

$$R = \frac{\eta e\lambda}{\hbar c} = \frac{\eta\lambda}{1.24} \tag{2.18b}$$

where R is in A/W and λ is μm. The quantum efficiency is dimensionless and is always less than unity. Thus R takes its maximum value when $\eta = 1$. While physicists prefer to use quantum efficiency, engineers often use responsivity; the two are related according to Equation 2.18. Detectors with a higher responsivity are preferred because they require less optical power.

Noise is unavoidable in a photodetector. As long as the signal is always much stronger than the noise, the effect of the latter is minimal. Noise sources exist in a photodiode and can degrade the performance of a receiver. Most noise sources are due to the random nature of the arrival of photons at the detector and the statistical nature of the generation of electron-hole pairs. The relationship between noise and signal is characterized by two terms: *signal-to-noise* ratio (S/N) and *bit-error rate* (BER). S/N denotes how much stronger the signal is than the noise, whereas BER represents the ratio of the number of bits decoded in error to the total number of bits received. BER replaces S/N for digital systems. In a fiber-optic system, S/N is regarded as the ratio of signal power to noise power at the input of the photodetector. The signal power is the power delivered to a hypothetical resistor R by the signal current I_s, while the noise power is the power delivered to the same resistor R by the noise current I_n. Thus,

$$\frac{S}{N} = \frac{P_s}{P_n} = \frac{\langle I_s \rangle R}{\langle I_n \rangle R} \tag{2.19a}$$

or

$$\frac{S}{N} = \frac{\langle I_s \rangle}{\langle I_n \rangle} \tag{2.19b}$$

where angle brackets ⟨ ⟩ denote an average over fluctuations.

Two major mechanisms are responsible for noise in a photodetector: shot noise and thermal noise. Shot noise is associated with the random process of the photon creating electron-hole pairs, which in turn generates the current

$$\langle I_{sn} \rangle = 2eI_o B \tag{2.20}$$

where B is the bandwidth of the electronics (such as a preamplifier) accepting the noise, and I_o is the average output current of the device. $I_o = I_d + I_p$, where I_d is the dark current and I_p is the photocurrent. The dark current is the current that continues to flow through the bias circuit when no light enters the photodiode as a result of electrons/holes that are thermally generated in the *pn* junction. Thermal noise is caused by the fluctuations in the load resistor of the photodetector. In general, any device or resistive load R_L will generate a thermal-noise mean-square current

$$\langle I_{tn} \rangle = \frac{4kTB}{R_L} \tag{2.21}$$

where T is the noise temperature of the device, $k = 1.38 \times 10^{-23}$ J/K is Boltzmann's constant, and B is as defined earlier.

2.3.2 *Avalanche photodiode (APD)*

A way of improving receiver sensitivity is to use an avalanche photodiode (APD), which provides current amplification within itself before the circuitry of the following amplifier. APDs are used as receivers for laser and ILD light sources. They are usually variations of PIN diodes. They are semiconductor junction detectors that have internal gain, which gives them higher responsivity than PIN detectors. Since they have larger values of responsivity R, they are used when the optical power available for the receiver is limited. They operate at high reverse bias voltage, typically above 300 V. This high applied voltage causes internal amplification or avalanche multiplication of the photocurrent. Under normal operation, the intrinsic layer is fully depleted. This state is known as a *reach-through* condition. For this reason, APDs are often known as *reach-through* APDs or RAPDs.

Figure 2.9 presents a typical APD. Light enters through the p region and is absorbed in the i region. The electrons that enter the p region are accelerated and impact other atoms, thereby creating more carriers. These carriers in turn create more carriers, leading eventually to the *avalanche effect*. This effect results in a multiplication of the current at the output terminals. The *multiplication factor* M is empirically given by

$$M = \frac{1}{1 - (V/V_b)^n} \tag{2.22}$$

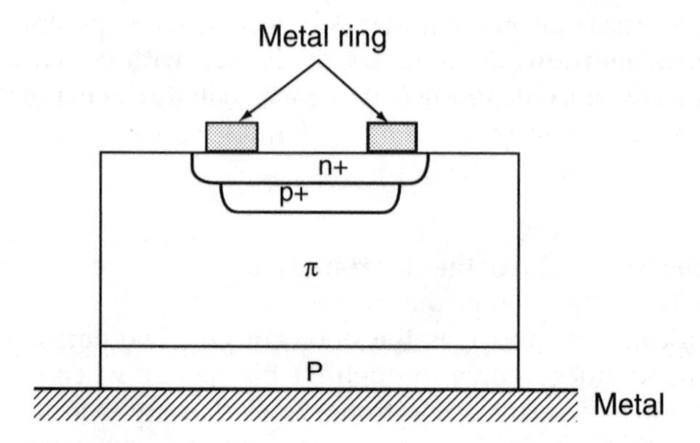

Figure 2.9 A typical APD.

where V is the applied reverse bias voltage, V_b is the breakdown voltage, and n is a constant between 2.5 and 7, depending on the material. The avalanche gain strongly depends on temperature and generally decreases as temperature rises. The APD is thus a more complex detector to make and use.

Just like the PIN diode, the performance of an APD is determined by its responsivity R. The current generated by an APD with gain M is

$$I_o = \frac{Me\eta P_o}{\hbar f} \tag{2.23}$$

Thus, the responsivity or sensitivity is

$$R = \frac{I_o}{P_o} = \frac{M\eta e\lambda}{\hbar c} = \frac{M\eta\lambda}{1.24} \tag{2.24}$$

where λ is in µm. Typical values of R for APD vary from 5 to 120 A/W.

In an APD, shot noise and thermal noise are still the two fundamental noise mechanisms. However, because of the multiplication process in an APD, the output noise is M^2 times the shot noise in Equation 2.20, where the average number of electrons is produced by a single primary electron. Also, a noise factor F is necessary to account for the random multiplication process causing a primary electron to produce more than M output electrons or sometimes less. Thus, Equation 2.20 becomes

$$\langle I_{sn} \rangle = 2eI_o BM^2 F \tag{2.25}$$

One may approximate F as

$$F(M) = M^x, \quad 0 \le x \le 1 \tag{2.26}$$

where x is an empirical constant that depends on the material. In addition to the dark current already mentioned, there is leakage current, which depends on the surface defects, surface area, and bias voltage. We need to include the noise due to this current. The mean square value of the noise due surface dark current is given by

$$\langle I_{sd}^2 \rangle = 2eI_L B \tag{2.27}$$

where I_L is the surface leakage current. From Equations 2.20, 2.21, 2.25, and 2.27, the signal-to-noise ratio is

$$\frac{S}{N} = \frac{\langle I_s^2 \rangle M^2}{2e(I_p + I_d)M^2 FB + 2eI_L B + 4kTB/R_L} \tag{2.28}$$

Equation (2.28) applies to both PIN and APD. For PIN diode, we set $M = 1$. APD optical receivers generally provide a higher S/N than PIN receivers for the same incident power because of the internal gain that increases the photocurrent by the multiplication factor M.

APD does have its problems. First, it has limited amplification. If too much amplification occurs, it will provide a continuous (large) current and cause a "run away" problem. Second, APD is noisy because thermal action can promote an electron to the conduction band. The current resulting from the thermal action is noise. Finally, the avalanche action takes time to occur, which means that the device has a certain response time.

2.4 Optical receivers

The primary purpose of an optical receiver is to detect the light incident on it and to convert it to an electrical signal. Hence, the receiver is an optical-to-electrical converter.

An optical receiver is usually a combination of photodetector, amplifier, and signal-processing elements, such as an equalizer and filter. These components may be on a single printed circuit board or an integrated circuit. Optical receivers can be part of the communication link or they can act as repeaters to extend the distance between terminals.

Typical block diagrams for digital and analog receivers are shown in Figure 2.10. Notice that the basic difference between two types of receivers is the way the signals are processed after amplification. Digital optical communications are more common than analog. Analog optical communication has found a useful primary application in cable television (CATV).

The optical detector converts the optical input into an electrical signal. The output current from a photodetector is usually very weak and needs extensive amplification to recover the information from it. Therefore, the

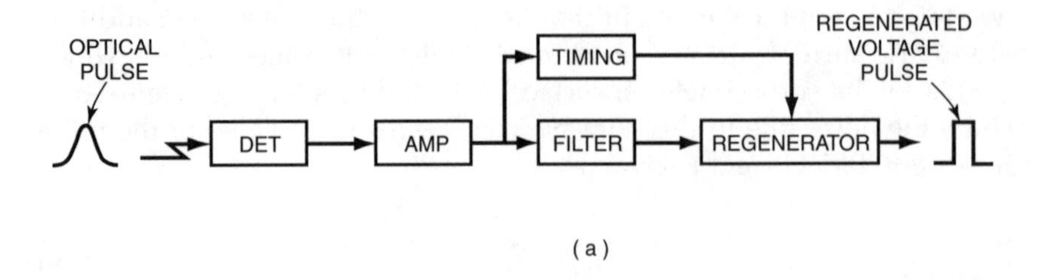

(a)

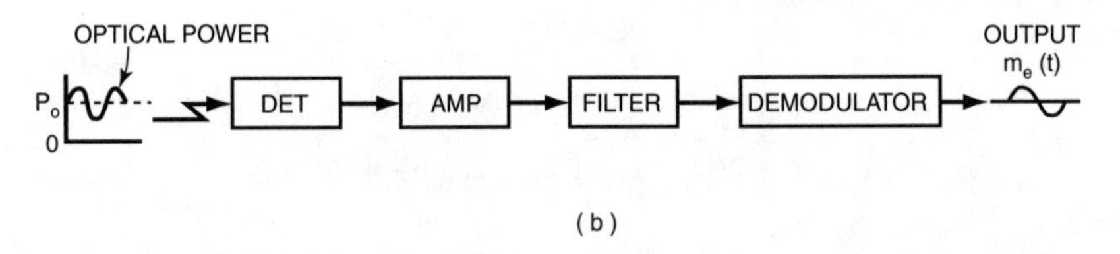

(b)

Figure 2.10 Typical receiver block diagram: (a) digital receiver, (b) analog receiver.

next stage is to amplify the signal to a predetermined reference for further processing. The amplification stage may require having the preamp, equalizer, and postamp. The equalizer that is between the two amplifiers works in conjunction with the preamp to restore the required bandwidth and to remove signal waveform distortions introduced by the amplifier and other components. The filter following the amplification stage is designed to remove unwanted frequency components generated up to this point. It also maximizes the S/N ratio while preserving the essential features of the signal. For a digital receiver, the signal feeds both a timing circuit and a digital signal regenerator. The timing circuit is basically a square-wave oscillator. It tells the regenerator when to sample, to ensure that the output information is clocked consistently, and ensures that all timing error (or jitter) is removed from the incoming pulse. The filter signal entering the regenerator is interpreted as 1 if high enough or 0 if low enough (below a threshold). For the analog receiver, the filtered signal is fed into a demodulator that separates the signal from the carrier.

A major aim in the design of an optical receiver is to minimize the power that must be regenerated by the receiver in order to maintain desired performance, measured in terms of the S/N ratio for analog systems or BER for digital systems. The basic performance parameters that determine the usefulness of the receiver are the wavelengths it will receive, sensitivity, and bandwidth. The sensitivity of the receiver is the minimum optical input power that is adequate to produce a useful output signal from the receiver. The bandwidth is primarily determined by the RC time constant of the amplifier input node and the filter or equalizer. A receiver must have high sensitivity, low noise, and high bandwidth.

For a digital receiver, one way of measuring the rate of error occurrence is using the bit error rate BER. This is the probability of incorrect identification of a bit by the decision circuit of the receiver. If N_e is the number of errors occurring in the time interval t, and N_t is the number of pulses (zeros and ones) transmitted within that interval, then

$$\text{BER} = \frac{N_e}{N_t} = \frac{N_e}{Bt} \tag{2.29}$$

where B is the bit rate. If three errors occur for every 1 million pulses transmitted, then BER = 3×10^{-6}. Typical values of BER range from 10^{-4} to 10^{-14}. It is standard practice in designing an optical link to specify the average optical input power to the receiver to achieve a prescribed BER, usually BER $\leq 10^{-9}$.

For analog receivers, the S/N is the major yardstick for determining the fidelity of the systems. From Equation 2.27, the S/N of the analog receiver is given by

$$\frac{S}{N} = \frac{\left(Me\eta P_o/\hbar f\right)^2}{2eI_o M^2 FB + 4kTB/R_l} \tag{2.30}$$

where $I_o = I_p + I_d$ is the output current and P_o is the average power of the light. In the denominator of Equation 2.30, only the shot noise and thermal noise are included since other noise sources do not contribute significantly. $M = 1$ for a PIN detector.

Summary

The semiconductor injection laser diodes (ILDs) provide a higher quality light source than light-emitting diodes (LEDs). ILDs are dominant in optical fiber communication systems although the LED is more economical. For higher data rates or with single-mode fibers, laser diodes are the optical source of choice. Caution: never look directly into either LED or ILD because the energy density of the emitted light is usually sufficient to damage the eye. The same is also true about looking into the end of an active optical fiber.

The detector of an optical fiber receiver will either be a PIN or avalanche photodiode. The PIN diode is cheaper and less sensitive to temperature, and it requires less bias voltage than the APD. The APD is more complex and has a more limited bandwidth than the PIN. The PIN diode is preferable in most systems.

Optical transmitters and receivers are designed for specific applications such as analog or digital. The role of the transmitter is to convert the input electrical signal into an optical signal and to launch the optical power from a light source into a fiber for transmission. A transmitter consists of electronic bias and an optical source. The task of the receiver is to convert the optical energy sensed by the photodetector into an electrical signal and then to amplify the signal to a level at which it can be processed by signal-processing circuitry. The receiver consists of a photo diode, an amplifier, and other support circuits. The performance of an optical receiver depends on the signal-to-noise ratio for an analog receiver or the bit error rate for a digital receiver. A more rigorous treatment of optical transmitter and receiver is found in Reference 5.

References

1. S. L. W. Meardon, *The Elements of Fiber Optics*, Prentice-Hall, Englewood Cliffs, NJ, 1993, 99–161.
2. M. J. N. Sibley, *Optical Communications*, McGraw-Hill, New York, 1990, 47–67.
3. W. F. Brinkman et al., The lasers behind the communications revolution, *Bell Labs Tech. J.*, vol. 5, no. 1, 2000, 150–167.
4. T. P. Lee and T. Li, Photodetectors, in S. E. Miller and A. G. Chynoweth (Eds.), *Optical Fiber Telecommunications*, Academic Press, Orlando, FL, 1979, 593–626.
5. G. P. Agrawal, *Fiber-Optic Communication Systems*, 2nd ed., John Wiley & Sons, New York, 1997, 75–192.

Problems

2.1 Explain the concepts of spontaneous emission, stimulated emission, and absorption.

2.2 An optical source is made of GaP, which has a bandgap energy of 2.26 eV. Calculate the frequency and wavelength of the optical carrier.

2.3 With the alloy fraction formulas in Equations 2.3 and 2.4, find the material composition for (a) a 1.3 μm source, (b) a 1.55 μm source.

2.4 At 300K, the quaternary alloy $Ga_{1-x}In_xAs_ySb_{1-y}$ has

$$E_g(x,y) = 0.726 - 0.961x - 0.501y + 0.08xy + 0.415x^2 + 1.2y^2 + 0.021x^2y - 0.62xy^2$$

(a) Show that E_g varies from 0.296 eV (4.2 μm) for $InAs_{0.91}Sb_{0.09}$ to 0.726 eV (1.7 μm) for GaSb.
(b) Find E_g and λ for $Ga_{0.65}In_{0.35}As_{0.15}Sb_{0.85}$.

2.5 The 3-dB frequency of an LED is desired to be 30 MHz. Find its modulation bandwidth.

2.6 Determine the cutoff wavelength of a photodiode made of GaAs.

2.7 Compare and contrast PIN photodiodes and avalanche photodiodes.

2.8 Define responsivity, quantum efficiency, and bit error rate.

2.9 A PIN diode has a quantum efficiency of 0.85 at 1.55 μm. Determine its responsivity.

2.10 For a PIN diode with quantum efficiencies of 0.5, 0.6, and 0.8, plot the responsivity vs. wavelength ($0.3 < \lambda < 1.5$ μm). Why does responsivity increase with wavelength?

2.11 A PIN photodetector operates with the following parameters:
Wavelength = 1.3 μm
Quantum efficiency = 65%
Temperature = 300 K
Load resistor = 20 kΩ
Input optical power = 65 nW
Noise factor = $M^{1/2}$
Bandwidth of receiver = 20 MHz
(a) Calculate the responsivity, (b) compare shot noise and thermal noise, and (c) find the signal-to-noise ratio.

2.12 Repeat the previous problem for an avalanche photodiode with a gain of 100; other parameters remain the same.

2.13 An APD with a gain of 80 has a quantum efficiency of 0.5 and a responsivity of 0.6 W/A. Find the operating wavelength and the photocurrent if the input power is 45 nW.

2.14 Calculate the responsivity of an InGaAs APD with a quantum efficiency of 0.8 and gain of 20 if it operates at 1.55 μm. How much optical power is required by this photodetector to produce 40 nA?

2.15 An APD has breakdown voltage of 350 V and a reserve voltage of 210 V. Calculate the multiplication factor when the material constant is 4.

2.16 Discuss the sources of noise in an optical receiver.

2.17 Consider an APD receiver. Using Equation 2.30 and the fact that $F = M^x$, derive an expression for the value of M for which the S/N becomes maximum.

chapter three

Optical multiplexers and amplifiers

> *Not everything that can be counted counts, and not everything that counts can be counted.*
>
> **— Albert Einstein**

The large bandwidth of fiber can be made full use of by transmitting several channels simultaneously on a single-mode fiber using multiplexing techniques. The multichannel lightwave systems that result can provide savings. The four major techniques used for optical signal multiplexing are (1) wavelength-division multiplexing (WDM) or optical frequency division multiplexing (OFDM), (2) optical time division multiplexing (OTDM), (3) subcarrier multiplexing (SCM), and (4) optical code division multiplexing (OCDM), also known as spread spectrum. Any of these techniques can be used in a multiple-access environment such as for broadcast in a LAN. A network that employs WDM is referred to as wavelength-division multiple access (WDMA) network. TDMA and CDMA networks are defined in a similar manner.[1]

This chapter presents these four types of multiplexing methods and two types of optical amplifiers. Although fiber has low loss, amplifiers are still required to regenerate the optical signal when transmitted over long distances.

3.1 WDM lightwave systems

In the past, light is transmitted through an optical fiber at a single wavelength (or color). In contrast, wavelength division multiplexing (WDM), or more generally optical frequency division multiplexing (OFDM), exploits the large potential bandwidth of the fiber by combining several wavelengths of light in a single optical fiber, as illustrated in Figure 3.1. A number of electrical sources each modulate a laser, all the lasers having different wavelengths. A multiplexer combines the wavelengths into a single fiber. At the receiving end of the fiber link, the different wavelengths are separated by a demultiplexer using

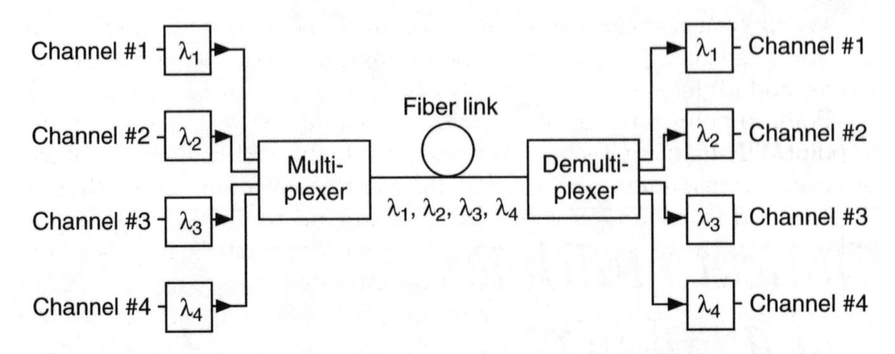

Figure 3.1 A four-channel point-to-point WDM transmission system.

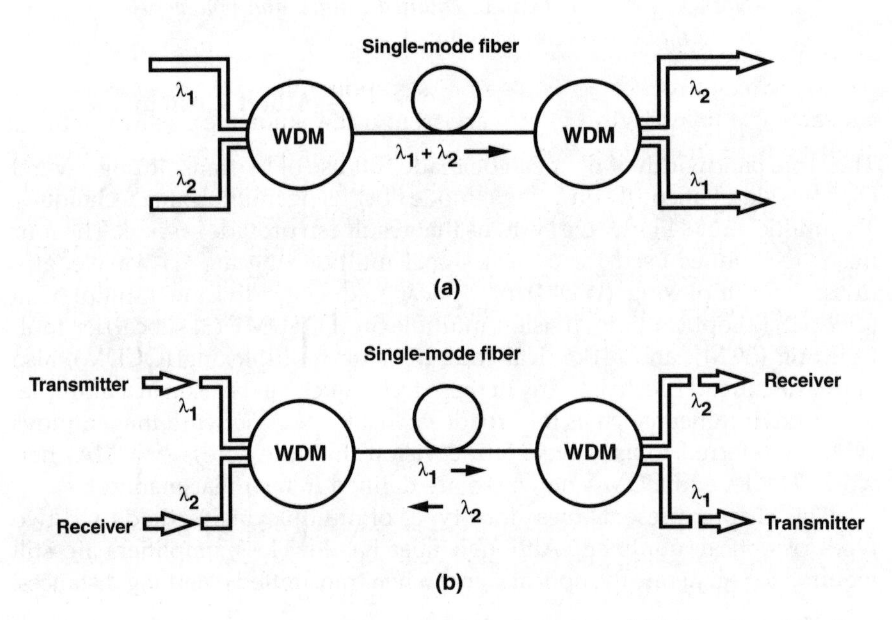

Figure 3.2 WDM systems: (a) unidirectional configuration, (b) bidirectional configuration.

a tunable optical filter. Individual detection of the channels is then performed. Figure 3.2 shows unidirectional and bidirectional WDM configurations.

It is common to classify and use different terms for WDM systems depending on the spacing between the channels. The wavelength spacing between the carriers or channels in WDM is on the order of 1 nm. In addition to WDM, there now are dense wavelength division multiplexing (DWDM) with channel spacing between 1 and 10 nm and high density wavelength division multiplexing (HDWDM) with channel spacing less than 1 nm. However, we will not make any distinction among these terms for now.

WDM systems can also be classified from the architectural viewpoint into three categories:[2,3] (1) point-to-point links, (2) broadcast-and-select networks, and (3) local area networks (LANs).

Point-to-point links are the simplest networks. An example of a point-to-point WDM system is shown in Figure 3.1. In this architecture, the output of several transmitters operating at different wavelengths is multiplexed together and launched into the optical fiber. At the receiving end, a demultiplexer separates the wavelengths and each carrier falls on a separate receiver for detection. If N channels transmit at bit rates B_1, B_2, ..., B_N are multiplexed over an optical fiber of length L, the total bit rate-distance product for the point-to-point link is

$$BL = \left(B_1 + B_2 + \cdots + B_N\right)L \tag{3.1}$$

If the bit rates are equal, the system capacity is increased by a factor of N. Thus, the major role of WDM in a point-to-point link is to increase the total bit rate. The link length L may vary from a few kilometers (short haul) to thousands of kilometers (long haul). If the length is beyond a particular value, it may be necessary to compensate for the fiber loss using regenerators (also called repeaters). A regenerator is essentially a receiver-transmitter pair that recovers the electrical bit stream from the incoming optical signal, amplifies it, and converts it back into optical form. The amount of light that leaks from one channel to another is known as *crosstalk*. Expressed in dB, the crosstalk XT from channel j into channel i is

$$\mathrm{XT}_{ij} = 10\log\left(\frac{P_{ij}}{P_{ii}}\right) \tag{3.2}$$

where P_{ii} is the power in channel i when only that channel is active and P_{ij} is the power in channel i when only channel j is active.

Point-to-point WDM transmission offers the following advantages:[4]

- Several signals can be carried by one fiber, thereby increasing system capacity with little complexity.
- System capacity can be easily upgraded.

It has the following disadvantages:

- Failure occurs when the cable is cut and the link goes down.
- Multiplexers/demultiplexers introduce loss.
- Individual repeaters or regenerators are needed for each wavelength.

The second type of WDM system is the broadcast-and-select network; a typical system is illustrated in Figure 3.3. In this architecture, information is

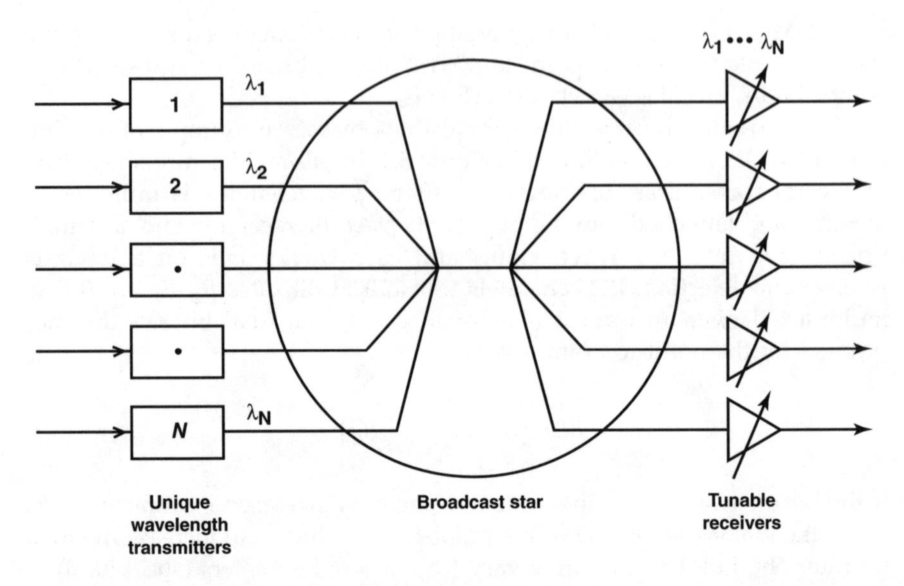

Figure 3.3 A typical broadcast-and-select WDM system.

not only transmitted but also distributed to a group of customers and the receiver selects each channel through multiplexing. Each customer receives all channels and selects one of them. Examples include the broadcast of multiple video channels over a cable television (CATV) and a wide distribution of services using a broadband integrated services digital network (BISDN). The throughput of a broadcast-and-select network is limited mainly by the distribution and insertion losses, which can be compensated by optical amplifiers.

The third type of WDM system is the multiple-access LANs, which provide bidirectional access to each customer. This offering of bidirectional access to customers makes multiple-access LANs different from broadcast-and-select networks. Each user can transmit to or receive from any other user in the network. The telephone network and Internet are examples of multiple-access LANs. But in the case of LANs, the transmission distance is short and the main motivation is exploiting the large bandwidth offered by the optical fiber.

The WDM LANs have several topologies (bus, ring, star, etc.). Figure 3.4 presents the star network as a typical WDM LAN, where each user has its own wavelength and the return fiber path is used by the user to select the appropriate wavelength from which to receive. The major advantage of LANs is that the transmission distance is short (less than 10 km) and fiber loss is not of much concern. Such networks are mainly limited by the electronics since each node must be able to process the traffic from the entire network.

The realization of WDM systems requires specific building blocks. WDM devices include transmitters, receivers, multiplexers, demultiplexers, tunable

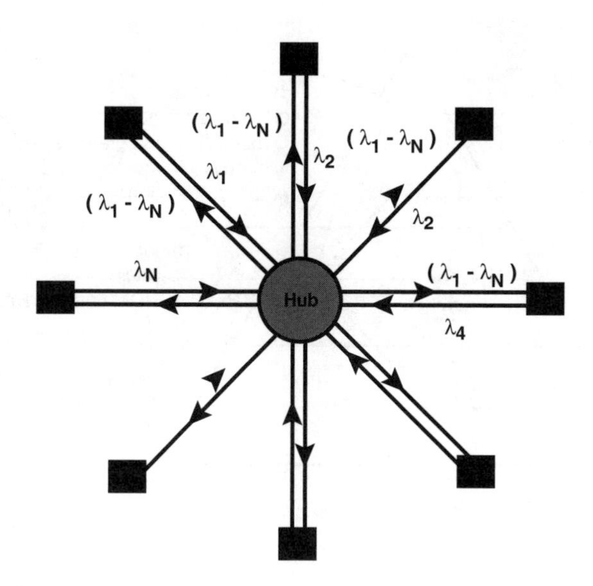

Figure 3.4 WDM star LAN.

optical filters, star couplers, routers, and cross-connects. A powerful source (LED or laser) with sufficient tenability to allow rapid selection of the desired wavelength is also required. As stated earlier, a multiplexer combines the output of several transmitters and launches it into an optical fiber. The demultiplexer splits the received signal into individual channels for different receivers. The same device can serve as a multiplexer or a demultiplexer, depending on the direction of propagation. An add/drop multiplexer is necessary when one or more channels are to be dropped or inserted. A star coupler combines the output of several transmitters and broadcasts the signal to different receivers. Star couplers do not contain wavelength-selective elements but do attempt to separate channels. The optical filter constitutes the basic building block of WDM systems because it provides the required wavelength selection to isolate individual channels. An optical filter selects or separates one channel at a specific wavelength at the receiver. It has a passband characteristic that can be changed by proper tuning. A wavelength router combines the functions of a star coupler with multiplexing and demultiplexing — the signal entering into an $N \times N$ router from each input port is split into N parts, one for each of the N output ports of the router. An optical cross-connect performs the same function as the digital switches in telephone networks.

WDM plays a major role in optical networking and is currently the main mechanism for realizing all-optical networks (AONs) where multiplexing is used to provide multiple access. Together with the newly developed erbium-doped fiber amplifiers (EDFAs), WDM systems have been largely responsible for another revolutionary generation of lightwave systems.

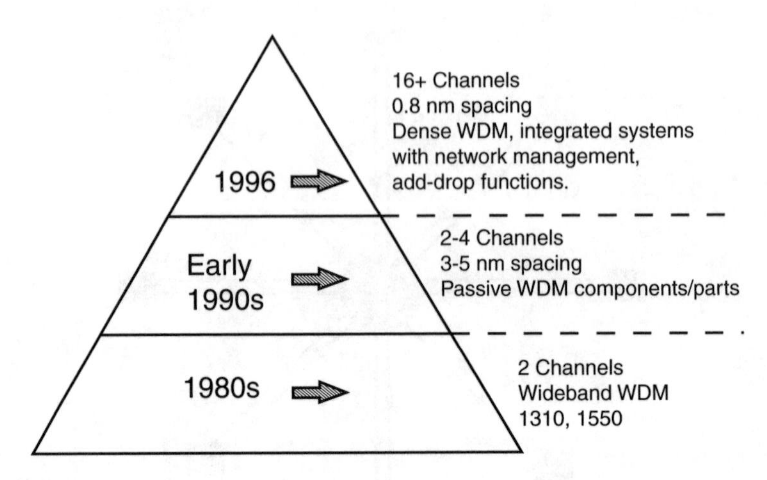

Figure 3.5 Evolution of WDM.

3.2 DWDM lightwave systems

The evolution of WDM is shown in Figure 3.5. Early WDM began in the late 1980s using wavelengths in the 1310 and 1550 nm regions. The early 1990s witnessed a second generation of WDM using 2 to 8 channels that were spaced at an interval of 400 GHz. By the mid-1990s, dense WDM (DWDM) systems emerged with 64 to 160 parallel channels, densely packed at 50 or 25 GHz intervals. The invention of DWDM was motivated by the need to increase the carrying capacity of optical fiber networks. (The principal driver, of course, is the seemingly inexhaustible human appetite for more bandwidth.) For example, the SONET/SDH rate of 39.81 Gbps (OC-768 or STM-256) was difficult to achieve for many years. Another example is the large amount of bandwidth required to carry digitally uncompressed analog video — 16 analog channels occupy an entire OC-48. Services such as video, high resolution graphics, and large volume data processing require unprecedented amounts of bandwidth. The growth of cellular and PCS is also placing more demand on fiber networks, which serve as the backbone even for wireless communications.

To meet the growing demands for bandwidth, a technology called dense wavelength division multiplexing (DWDM) has been developed. DWDM multiplies the capacity of a single fiber. It allows the transport of 80 analog channels, while maintaining 5 gigabits for voice and data services.

The emergence of DWDM is one of the most recent and important phenomena in the development of optical transmission technology. DWDM increases the capacity of existing and new fiber by assigning incoming optical signals to specific frequencies or wavelengths (called lambdas) and then multiplexing the resulting signals on a single fiber. To transmit 40 Gbps over 600 km using a traditional system would require 16 separate fiber pairs with

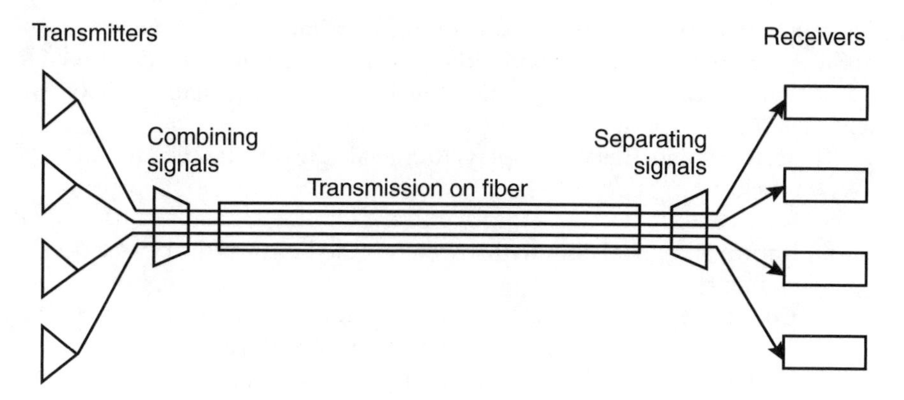

Figure 3.6 DWDM functional schematic.

regenerators placed every 35 km for a total of 272 regenerators. For the same transmission, a 16-channel DWDM system, on the other hand, uses a single fiber pair and four amplifiers positioned every 120 km. Using DWDM, single fibers have been able to transmit at speeds up to 400 Gbps, which is sufficient to transmit 90,000 volumes of an encyclopedia in 1 s. Terabit capacity is on its way.

A DWDM system is capable of transmitting up to 16 discrete optical channels over one fiber pair. Each channel is bit-rate transparent from 150 Mbps to 2.4 Gbps and operates with existing SONET/SDH terminals.

DWDM system basically consists of transmitters, receivers, Erbium-doped fiber amplifiers (EDFA), DWDM multiplexors, and DWDM demultiplexors. It involves a number of physical-layer functions; a typical configuration is depicted in Figure 3.6. Figure 3.6 also shows a DWDM schematic for four channels with each channel occupying its own wavelength. The DWDM system performs the following functions:

- *Generating the signal:* The source, a solid-state narrowband laser with precise, stable wavelength, provides light within a specific bandwidth to carry the digital data.
- *Combining the signals:* Optical add/drop multiplexers are employed to combine the signals.
- *Transmitting the signals:* An optical fiber with low loss transmits data parallel-by-bit or serial-by-character. The effects of crosstalk and attenuation must be reckoned with. On the transmission link, the signal may need to be optically amplified. The invention of the flat-gain optical amplifier, coupled in line with the transmitting fiber to boost the optical signal, dramatically increases the viability of DWDM systems by greatly extending the transmission distance.
- *Separating the received signals:* The multiplexed signals must be separated at the receiving end.
- *Receiving the signals:* A photodetector receives the demultiplexed signal.

The most common form of DWDM employs a fiber pair — one for transmission and one for reception. Although a single fiber can be used for bidirectional traffic, this configuration must sacrifice some fiber capacity and degrade amplifier performance.

The technologies that have played crucial roles in the development of DWDM include:

- Improved optical fiber with lower loss and better optical transmission characteristics.
- Demultiplexers that use fiber Bragg grating: A fiber grating is a filter component that consists of a length of optical fiber wherein the refractive index of the core has been permanently modified in a periodic fashion. It is a passive device, fabricated into glass fiber. The fiber grating creates a highly selective, narrow bandwidth filter and provides significantly greater wavelength selectivity than any other optical technology.
- Erbium-doped fiber amplifier (EDFA): EDFA provides a way of amplifying all the wavelengths simultaneously. Instead of multiple electronic regenerators converting the optical signals to electrical signals and then back to optical ones, the EDFA directly amplifies the optical signals. Thus, the composite optical signals can travel up to 600 km without regeneration. More will be said about this amplifier later.

The ability to harness the potential of these enabling technologies led to the realization of DWDM systems.

One of the main applications of DWDM is long-distance telecommunications where operators can use either point-to-point or ring topologies. By deploying DWDM terminals, an operator can construct a 40 Gbps self-healing ring, with 16 separate communication signals using only two fibers. The transparency of DWDM systems to various bit rates and protocols will allow carriers to tailor and segregate services to diverse customers along the same transmission routes. Another application of DWDM is found in high-speed MAN architectures typically used in access networks.

DWDM-based optical networks provide near-limitless amounts of bandwidth capacity. An additional benefit of DWDM is that it is protocol and bit-rate independent. That is, DWDM provides a "highway" that is indifferent to the type of traffic traveling on it. Thus, DWDM-based networks can carry different types of traffic at different speeds over an optical channel and can transmit data in IP, ATM, SONET/SDH, and Gigabit Ethernet. Each signal carried can be at a different rate (OC-3,-12,-24, etc.) and different format (IP, ATM, etc.). Furthermore, a DWDM infrastructure also increases the distances between network elements (compared with repeater-based applications) — a significant benefit for long-distance service providers. Using fewer regenerators in long-distance networks results in fewer interruptions and improved efficiency.

The other side of the coin is that replacing an existing single-mode fiber system with DWDM requires the replacement of the electronics, an expensive process that causes an interruption of service to the customers.

3.3 OTDM lightwave systems

Time-division multiplexing (TDM) is a popular technique in electronic communications. It has been overwhelmingly used in point-to-point digital data transmission and is the predominant scheme for transmitting multiple channels of PCM analog information, such as voice or video. Its broadcast network version, *time-division multiple access* (TDMA), has been used widely in satellite communication networks. However, implementing TDM becomes difficult when the bit rate is above 10 Gbps because of the imposed limitation of the electronic circuitry. A solution is offered by the optical TDM (OTDM), which is aimed at increasing the bit rate to as high as 1 Tbps.

TDM transmits several digital signals over an optical fiber on a time-sharing basis. The fiber carries many time-multiplexed channels, in which each channel transmits its information in a preassigned time slot. The time slot identifies the sender's (or receiver's) address. No contention problem exists with TDM since the scheme allows only one user to transmit in a given slot. A typical TDM link is illustrated in Figure 3.7, in which four slow-speed bit streams are polled and merged into a single high-speed stream. Each input stream in turn is assigned one bit in every four. Each signal must be synchronized to the higher-speed stream. The time-multiplexed data is transmitted along the fiber and demultiplexed at the receiving end. Nodes in a TDM system can be either synchronous or asynchronous. In a synchronous TDM system, a node gets the time allocated for it even if it is idle or has nothing to transmit. On the other hand, in an asynchronous system, only active nodes are given access to the channel although all the nodes are still polled in a round-robin fashion. The most common TDM network is the telephone network.

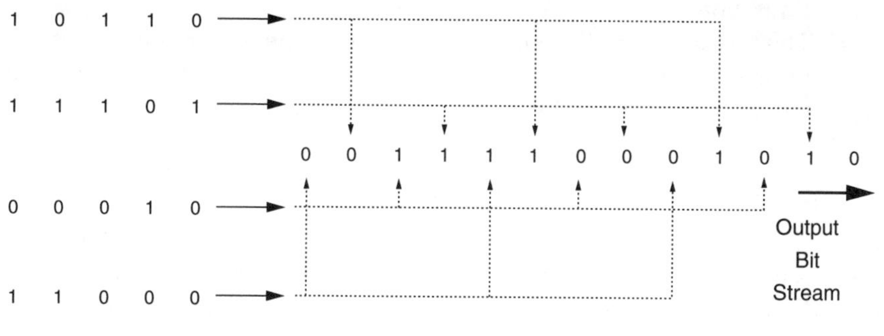

Figure 3.7 Illustration of the typical TDM concept.

The major difference between TDM and OTDM lies in the use of multiplexing and demultiplexing in the optical domain. OTDM systems are generally synchronous because of their high data rate. OTDM is based almost exclusively on slot or bit interleaving (instead of byte interleaving used in SDH or 53-byte cell interleaving used in ATM) of N independent optical streams at B Gbps. The data stream is arranged in repeating patterns of time slots, called *frames*. In the example in Figure 3.7, a frame consists of four bits. In conformity with the RZ (return-to-zero) code suitable for subsequent interleaving, each bit time is divided into two halves. For a 1-bit, the first half of the bit time is occupied by an optical pulse while the second half is dark. For a 0-bit, the whole bit time is dark.

The OTDM transmitter requires an optical source (laser) that can generate a periodic pulse train at a rate equal to the single-channel bit rate B and produce pulse width $T < 1/(NB)$ to guarantee that the pulse will not overlap with other adjacent pulses. Demultiplexing of an OTDM stream demands the recovery of a clock signal at the bit rate of a single channel. This requires an all-optical scheme because of the high bit rate involved in OTDM. The most practical demultiplexers are based on electro-optic switches.

OTDM has been used in designing optical networks that can connect multiple nodes of random bidirectional access. Both single-hop and multihop architectures have been considered. The main advantages of OTDM include:

- There is no contention problem because each bit occupies its own time slot.
- Electronic bottlenecks are avoided in the transmission path.
- The implementation of the appropriate photonic networks is straightforward and similar to electronic networks.

The major difficulties include:

- High-speed implementation is difficult.
- The scheme requires ultrahigh speed switches that have bandwidth limitations.
- The transmission distance is limited due to the required transmission of short pulses, which are prone to fiber dispersion and nonlinear effects.

3.4 SCM lightwave systems

The idea of subcarrier multiplexing (SCM) came from microwave communications that apply multiple microwave carriers to transmit multiple channels over coaxial cables or free space. It essentially performs modulation/demodulation, multiplexing/demultiplexing, and electrical rather than optical routing. Of course, other functions, such as optical generation, optical detection, and optical coupling, remain.

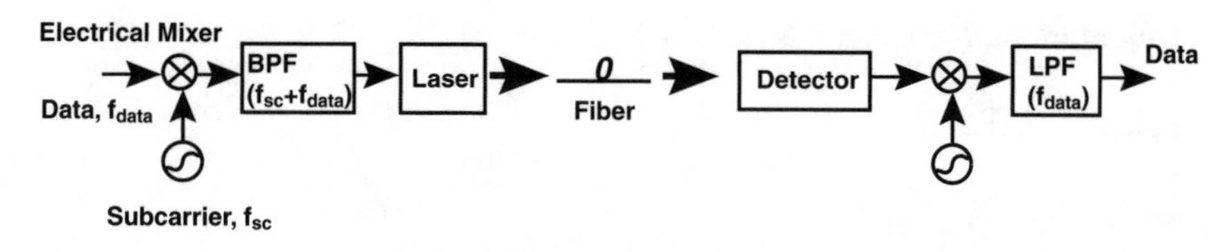

Figure 3.8 A typical lightwave SCM system.

This partial replacement of optics with electronics is the major advantage of SCM. Analog optical communication is appealing because of its compatibility with much of the analog transmission in use today and because of its commercial benefit for transmitting a large number of channels for broadband services. Such services include CATV, video telephony, interactive video, high definition television, personal communication systems, and LANs with bit rates greater than 1 Gbps.

A typical SCM system is shown in Figure 3.8. In this scheme, each wavelength channel is partitioned into nonoverlapping subchannels or subcarrier frequencies. A number of channels are first multiplexed in the electrical domain in the microwave frequency range (from 0.3 to 30 GHz), and the output is used to modulate a single laser (optical transmitter) at a particular wavelength. The microwave carrier is known as the *subcarrier,* while the optical carrier is regarded as the main carrier. This technique is known as *subcarrier modulation*. For instance, with 50 wavelengths and 20 subcarrier channels to each wavelength, it is possible to transmit a total of 1000 channels over a single fiber.

Subcarrier schemes can be used by themselves or in combination with other multiplexing schemes such as TDM or WDM. For example, combining SCM and WDM can achieve a bandwidth in the range of 1 THz. The application of WDM and SCM for personal communications networks is receiving some attention. SCM is used today by cable operators to transmit multiple analog video signals using a single optical transmitter.

The SCM techniques offer some advantages over WDM schemes. First, both analog and digital signals can be transmitted simultaneously in one system. Second, SCM offers low cost by using commercially available electronics. Third, SCM networks offer broadband services to a large number of subscribers from a single transmission point.

3.5 CDM lightwave systems

Code-division multiplexing (CDM) is a technique that allows users to share the same channel by transmitting different types of signal encoding at the same wavelength. It is based on the concept of spread spectrum, in which a modulation process is used to spread the encoded signal spectrum over a much wider bandwidth than that of the original signal. The spread spectrum allows many simultaneous channels to be accommodated on a shared

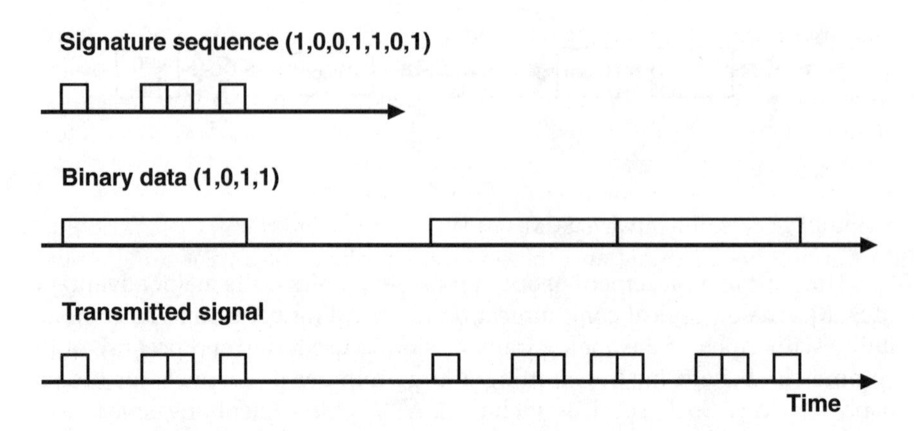

Figure 3.9 Direct-sequence coding scheme for CDM system.

medium with a minimal coordination. It is also immune to noise and interference. It has been widely used in mobile-satellite and digital-cellular communication systems.

For multiple users to share a communication channel with a fixed amount of bandwidth, a multiple-access technique is needed. In WDM, the channel bandwidth is divided into small wavelength bands, whereas in TDM time is divided into slots. In CDM, all users are permitted to access simultaneously the entire channel bandwidth. Also, in CDM there is no limit to the number of users, whereas WDM has a fixed number of channels and TDM has a fixed number of slots.

The basic concept of CDM is to have the transmitter encode each bit of a data stream with a unique waveform. The unique code is independent of the data itself and is similar to a distinct wavelength in WDM or a specific time slot in TDM. Each user is assigned a unique code that is used to encode data sent by that user. A receiver will receive only the code to which it is tuned. The receiver utilizes the same unique code for recovering the data and filters out other waveforms.

There are several methods of coding. Two popular methods are direct sequence and frequency hopping. In a direct-sequence coding scheme, each bit of data is with a signature sequence or chip. If the sequence has length L, each data bit is represented by L chips. Thus, the effective transmission rate is increased by a factor of L. A typical example is shown in Figure 3.9, where the signature sequence used for the coding has seven bits (1, 0, 0, 1, 1, 0, 1). Notice that the sequence is fixed and would be the same for every 1-bit transmitted; 0-bit is not transmitted at all. Each user is assigned a unique signature sequence, and each receiver recovers that data by applying the corresponding signature sequence used to code the message.

In frequency hopping, the waveform is generated as a sinusoidal carrier with a frequency that is varied (or hopped) in a pseudorandom manner over the bit duration. The bandwidth is segmented into frequency channels and the transmitted signals hop from one frequency to another. Each user is assigned different frequency-hop codes to make sure that two users do not transmit at the same frequency during the same time slot. For example, a user may transmit six data bits on frequency channel f1, hop to another frequency f2, transmit another six data bits, and so on. For either direct sequence or frequency hopping, the process of creating a CDM waveform may be performed either electronically or optically.

CDM has some advantages over WDM and TDM. First, there is no limit to the number of users, and there is the simplicity of data routing among users. Second, unlike in WDM or TDM, its bit rate is not fixed. Third, it provides increased transmission security in that only receivers with the appropriate, unique code can receive a given message; CDM makes jamming a signal difficult because of its coded nature. Fourth, it is possible to overlay onto frequency bands occupied by narrowband users.

CDM has its own drawbacks. First, the power from distant transmitters is less than the power from closer transmitters, thereby resulting in high degradation for the distant transmitters. Second, there is a limitation in the speed of the system. If the original bit rate is high, physical devices may not be able to handle a much higher chip rate, which is L times the original bit rate. This is the most crucial disadvantage at the moment because gigabit transmission can be achieved in CDM only with great difficulty. Third, there are high splitting losses at the encoder/decoder.

3.6 Optical amplifiers

The advent of optical amplifiers has revolutionized communications. These devices have enabled new and exciting optical systems to be conceived and demonstrated. In fact, optical amplifiers have been the key network components in most recent advances in optical communications.

Optical amplifiers were relatively unknown before 1980. Traditionally, when the optical power in a link reaches the minimum detectable level, a repeater is added. The repeater performs photon-to-electron conversion, electrical amplification, pulse shaping, and finally electron-to-photon conversion. This process becomes cumbersome and expensive for high-speed multiwavelength systems. This situation motivated a great deal of effort in all-optical amplifiers, which operate completely in the optical domain. Because of fiber attenuation, wide area all-optical networks cannot exist without optical amplifiers.

An optical amplifier is a device that amplifies the optical signal directly without photon-to-electron conversion; it amplifies the light itself. The ideal amplifier would have the following properties:[5]

- Provide high gain and uniform gain over its spectral width
- Have broad bandwidth
- Allow bidirectional operation
- Have low insertion when inserted into a fiber link
- Have good conversion efficiency

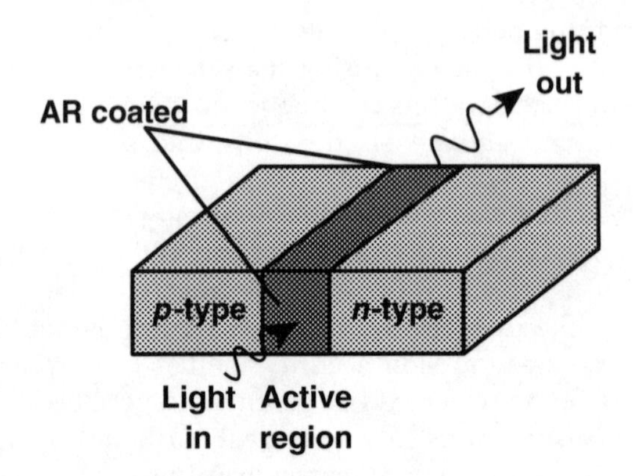

Figure 3.10 Semiconductor optical amplifiers.

There are essentially two types of amplifiers used in communications: semiconductor laser amplifiers (SLAs) and erbium-doped fiber amplifiers (EDFAs). Although SLAs are not as good as EDFAs for use as amplifiers, the former are finding applications in switches and wavelength-converting devices for deployment in optical networks.

3.6.1 Semiconductor amplifiers

Semiconductor optical amplifier is essentially a semiconductor laser with or without facet reflections. It is a pn-junction with a thin layer of a different semiconductor material sandwiched between the *p*-type and *n*-type regions, as shown in Figure 3.10. The semiconductor material forms the active region. The material used for the active region has a higher refractive index to help confine the light during amplification. Amplification takes place when light propagates through the active region. Like laser diodes, a positive gain in semiconductor amplifiers comes from external current injection. The amplifier gain is defined as

$$G = \frac{P_{out}}{P_{in}} \tag{3.3}$$

where P_{in} and P_{out} are, respectively, the input and output powers of the optical signal being amplified. If G_o is the single-pass gain in the absence of light, the amplifier gain G is given by the transcendental equation

$$G = G_o \exp\left(-\frac{(G-1)}{G}\frac{P_{sat}}{P_{in}}\right) \tag{3.4}$$

where P_{sat} is the amplifier saturation power or the internal power level at which the gain per unit length has become half.

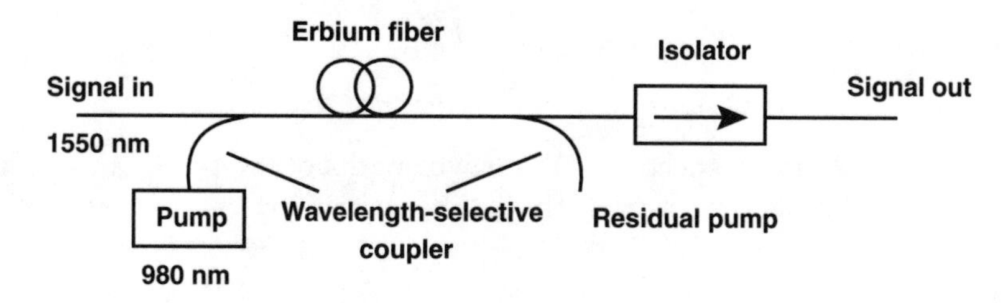

Figure 3.11 EDFA optical amplifiers.

3.6.2 Erbium-doped fiber amplifiers

The erbium-doped fiber amplifier (EDFA) has generated much excitement in optical communications primarily because it has many desirable features such as:[6]

- The device is simple.
- It is an all-fiber device, making it virtually polarization insensitive.
- It operates near 1550 nm wavelength region where fiber loss is minimum.
- It introduces no crosstalk and low distortion when amplifying WDM signals.
- Erbium produces amplification in a fairly wide band.
- Compact high-power semiconductor efficient pump lasers are available.
- It has high output power and very high sensitivity.

The EDFA belongs to a class of earth-doped amplifiers. The active medium in the amplifier is made of a 10- to 30-m length of optical fiber lightly doped with a rare-earth element such as erbium (Er), neodymium (Nd), ytterbium (Yb), or praseodymium (Pr). The most widely used material for long-distance communications is a silica fiber doped with erbium.

As shown in Figure 3.11, the EDFA consists of a section of doped single-mode fiber, one or more pump lasers to provide the energy necessary for population inversion, a passive wavelength coupler to combine signal and pump, and optical isolators to prevent reflections returning from the fiber.

The principle used here is that of a laser, and it is simple. Erbium ions can exist in several energy states. When in a high-energy state, an erbium ion is stimulated by a photon of light to give up some of its energy and return to a lower-energy state, a form of stimulated emission. However, to excite erbium ions to a higher energy state, external optical pumping through a directional coupler is needed. The external pumping must be operated at a higher frequency than that of the amplified signal. A pumping source at wavelength 1480 nm or 980 nm (a wavelength matched to the characteristic of erbium) can be used. The pump is coupled into the transmission fiber through a wavelength-selective coupler. If the input signal power is $P_{s,in}$ and the signal output power is $P_{s,out}$, by the principle of energy conservation,

$$P_{s,out} \leq P_{s,in} + \frac{\lambda_p}{\lambda_s} P_{p,in} \tag{3.5}$$

where λ_p and λ_s are, respectively, the wavelengths of the pump and signal, and $P_{p,in}$ is the input pump power. This can be written in terms of the gain G as

$$G = \frac{P_{s,out}}{P_{s,in}} \leq 1 + \frac{\lambda_p}{\lambda_s} \frac{P_{p,in}}{P_{s,in}} \tag{3.6}$$

Among the disadvantages of EDFAs are:

- They are limited to operate near 1550 nm.
- They require a considerable amount of pumping power (40 to 50 mW) at 1480 nm.
- They require lengths of fiber.
- They add noise to the optical signal.

Despite these disadvantages, EDFAs are used routinely for terrestrial WDM systems, wide area networks, undersea links, and video distribution.

A new type of amplifier, known as a Raman amplifier, has recently been invented. Raman amplification exploits stimulated Raman scattering (SRS). Since SRS occurs in regular fiber, not requiring any doping, Raman amplification can be added to existing installed fiber. With the right choice of wavelength for the pump signal and the information signal, power is transferred from the pump signal to the information signal. However, the gain provided by current Raman amplifiers is modest and does not eliminate the need for EDFAs. Since Raman amplification can be done from existing EDFA sites by inserting a pump at the proper place, it can allow existing fibers to be used for OC-768 without requiring additional EDFAs.

Summary

This chapter has examined several multiplexing schemes and optical amplifiers:

- WDM is a medium-access technique for transmitting traffic over fiber in multiple channels. It is the operation of multiple optical links over one fiber.
- DWDM is an optical technology used to increase bandwidth over existing and new fiber-optic backbones. It is already established as the preferred architecture for relieving the bandwidth crunch many carriers face.
- TDM is a technique that assigns a certain time slot in which a signal can be put on a single channel.
- OTDM is designed to avoid the electronic bottleneck currently around 10 Gbps.

- SCM is a scheme in which an additional level of modulation/demodulation is introduced electronically by using microwave subcarriers.
- CDM is a multiplexing technique that makes use of spread spectrum.

The technologies for OTDM and CDM are less mature than WDM, but discussion in this chapter would be incomplete without treating OTDM and CDM. WDM will enable telephone networks to cope with the increasing bandwidth demands. It is expected that the next generation of Internet will use WDM-based optical backbones. Optical communication technology is currently moving from point-to-point systems to optical networking. Optical amplifiers will play a major role in the global next generation optical networks.

References

1. B. Mukherjee, *Optical Communication Networks,* McGraw-Hill, New York, 1997, 5.
2. G. P. Agrawal, *Fiber-Optic Communication Systems,* 2nd ed., John Wiley & Sons, New York, 1997, 193–199, 284–360.
3. R. Papannareddy, *Introduction to Lightwave Communication Systems,* Artech House, Boston, 1997, 209–234.
4. D. M. Spirit and M. J. O'Mahony, *High Capacity Optical Transmission Explained,* John Wiley & Sons, Chichester, 1995, 197–216.
5. J. P. Powers, *An Introduction to Fiber Optic Systems,* Irwin, Boston, 1993, 293–307.
6. R. Ramaswami and K. N. Sivarajan, *Optical Networks: A Practical Perspective,* San Francisco, 1998, 122–123.

Problems

3.1 Why is WDM currently the most popular choice for optical communication networks?

3.2 Discuss the merits and drawbacks of DWDM-based networks.

3.3 What are the enabling technologies for the realization of DWDM?

3.4 Confronted by the demand for more bandwidth, what choices do you have?

3.5 Discuss the difference between WDM and TDM.

3.6 Explain the SCM scheme.

3.7 A receiver has a bandwidth of 12 GHz. Determine how many SCM channels can be recovered if each channel is amplitude modulated at 100 MHz. Assume that the spacing between channels must be twice the modulation rate.

3.8 What makes CDM attractive?

3.9 A direct sequence is used in a CDM system in which three users share a channel. The users transmit as follows:

User 1 transmits data D1 = 011 using signature sequence SS1 = 1100101
User 2 transmits data D2 = 101 using signature sequence SS2 = 1010011
User 3 transmits data D3 = 100 using signature sequence SS3 = 0110101

Show the coded waveform for each case.

3.10 Show that Equation 3.4 can be written as

$$G = 1 + \frac{P_{sat}}{P_{in}} \ln\left(\frac{G_o}{G}\right)$$

3.11 An EDFA is pumped at 980 nm with a 40-mW pump. If the gain at 1480 nm is 20 dB (or 100), find the maximum input power.

chapter four

Optical networks

I cried because I had no money until I saw a man that had no sense.

— Jackson T. Cassady

Thus far in this book, the basic technologies for optical networks have been examined. This chapter discusses their applications in the networks. The fact that next generation communication networks face some challenging problems should be kept in mind. Such problems include:

- Support of integrated service (voice, data, and video)
- Real-time traffic support
- Ability to handle exciting new applications such as desktop videoconferencing, interactive TV, supercomputer interconnection, and telemedicine
- Growth flexibility and fault tolerance
- Very high bandwidth requirements at low cost

To meet these and other requirements, several optical network proposals have been made. These networks can be roughly categorized as:

- Local area networks (LANs), which interconnect users within a small geographical area, such as a department, building, or campus
- Metropolitan area networks (MANs), which interconnect users within a city or a metropolitan area
- Wide area networks (WANs), which interconnect users within a large geographical area such as a country or the globe

The first generation optical networks merely use optical fiber as a transmission medium to replace copper cables. They allow electronic devices to handle switching and routing. This chapter will first introduce such optical networks, which include fiber channel and fiber distributed data interface

(FDDI) used as LAN, and synchronous optical network (SONET) deployed as a WAN.

To increase the capacity on existing fiber, wavelength division multiplexing (WDM) is used. WDM transmits data simultaneously on multiple carrier wavelengths over the same fiber. Today, WDM is mainly used in broadcast-and-select networks in the LAN arena and wavelength-routed networks as WANs. These two types of WDM networks along with undersea networks will be discussed. In addition, this chapter will discuss other emerging technologies relating to optical networking, such as optical Gigabit Ethernet, dynamic synchronous transfer mode (DTM), and multiprotocol label switching (MPLS).

4.1 FDDI networks

Fiber distributed data interface (FDDI) is a collection of standards formed by the ANSI X3T9.5 task group over a period of 10 years. The standards produced by the task group cover physical hardware, physical and data link protocol layers, and a conformance testing standard.[1–12] The original standard, known as FDDI-I, provides the basic data-only operation. An extended standard, FDDI-II, supports hybrid data and real-time applications.

4.1.1 Basic features

FDDI is both an interface technology and a protocol. The FDDI specification recommends an optical fiber with a core diameter of 62.5 μ and a cladding diameter of 125 μ. The two main attributes of FDDI are its size and transmission rates. It operates at 100 Mbps and can connect as many as 500 nodes per ring, with up to 2 km between nodes and a total circumference of about 100 km. The FDDI is therefore classified as a MAN as well as a LAN. The key highlights of FDDI are as follows:

- Dual counter-rotating ring topology for fault tolerance
- Data rate of 100 Mbps
- Total ring loop of size 100 km
- Maximum of 500 directly attached stations or devices
- 2 km maximum distance between stations
- Variable packet size (4500 bytes, maximum)
- 4B/5B data encoding scheme to ensure data integrity
- Shared medium using a timed-token protocol
- Variety of physical media, including fiber and twisted pair
- 62.5/125 microns multimode fiber-optic based network
- Low bit error rate of 10^{-9} (1 in 1 billion)
- Compatibility with IEEE 802 LANs by use of 802.2 LLC
- Distributed clocking to support large number of stations
- Support for both synchronous and asynchronous services

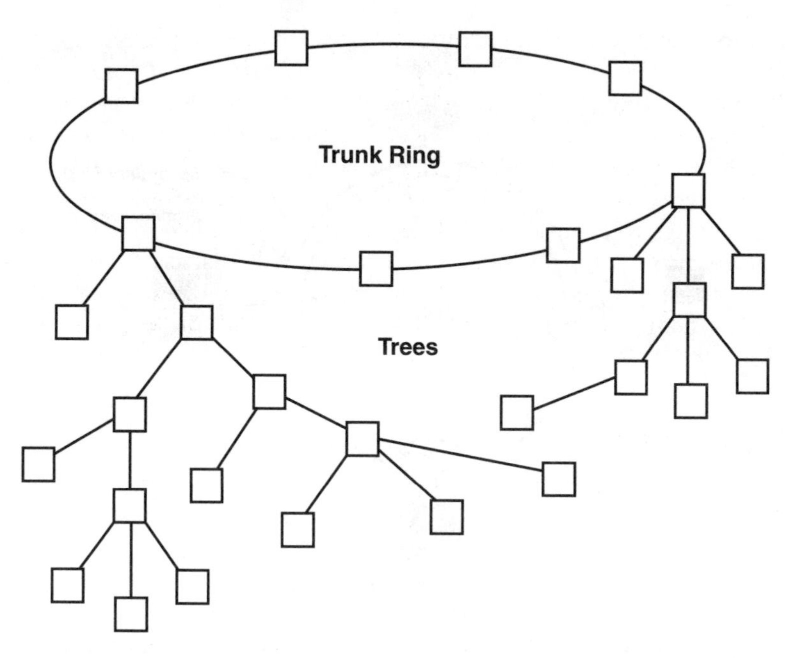

Figure 4.1 FDDI ring-of-trees topology.

FDDI has two types of nodes: stations and concentrators. Stations transmit information to other stations on the ring and receive from them. Concentrators are nodes that provide additional ports for attachments of stations to the network. A concentrator receives data from the ring and forwards it to each of the connected ports sequentially at 100 Mbps. While a station may have one or more MACs, a concentrator might not have a MAC.

There are two types of stations: single-attachment stations (SASs) and dual-attachment stations (DASs). A single-attachment station connects to only one ring through a concentrator. A dual-attachment station is connected to both rings and has two ports.

Figure 4.1 shows the most general FDDI topology (known as *ring of trees*), where each tree denotes a subsection of the network. In this case, the dual ring serves as the trunk ring for a complex topology. FDDI can be configured into a variety of topologies including dual ring with or without trees, wrapped ring with or without trees, and single tree. In each of these topologies, FDDI supports two data paths: a primary data path and a secondary data path. As shown in Figure 4.2, each FDDI station is connected to two rings, primary and secondary, simultaneously. Stations have active taps on the ring and operate as repeaters, which allows FDDI networks to be so large without signal degradation. The network uses its primary ring for data transmission, while the secondary ring can be used either for data or to ensure fault tolerance. When a station or link fails, the primary and secondary rings form a single one-way ring, isolating the fault while maintaining a logical

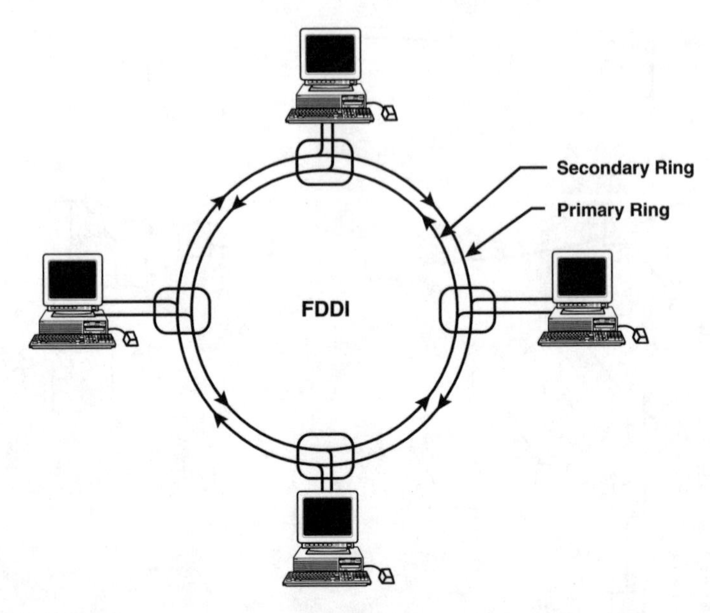

Figure 4.2 FDDI rings.

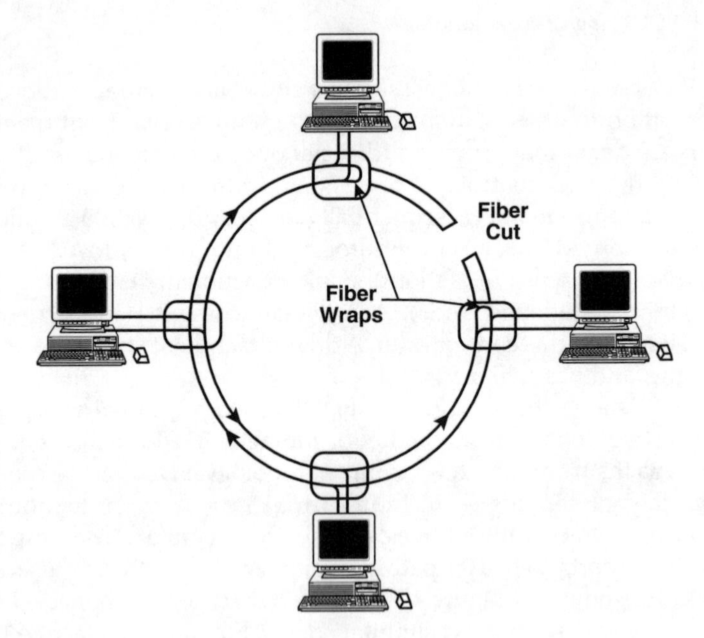

Figure 4.3 FDDI isolates fault without impairing the network.

path among users, as shown in Figure 4.3. Thus, FDDI's dual-ring topology and connection management functions establish a fault-tolerance mechanism.

FDDI was developed to conform with the OSI reference model. It divides the physical layer of the OSI reference model into two sublayers: physical medium dependent (PMD) and physical layer (PHY), while the data link layer is split into two sublayers: media access control (MAC) and IEEE 802.2 logical link control (LLC). Thus, architecturally, an FDDI station consists of four standardized layers:

- *PMD:* Physical medium dependent (or the media layer) describes the electrical or optical link connection to the FDDI ring. PMD options include:
  - multimode fiber (MMF-PMD)
  - single-mode fiber (SMF-PMD) and low-cost fiber (LCF-PMD) (which allows low-cost transmitters and receivers to be used on a FDDI link but the maximum distance between nodes is 500 m instead of the original 2 km)
  - twisted-pair (TP-PMD)
  - both shielded (STP) and unshielded (UTP) copper cable
  - synchronous optical network (SONET) physical layer mapping (SPM)

  As mentioned earlier, a dual-attached multimode fiber FDDI ring can run up to 200 km in length with stations 2 km apart, without requiring repeaters to boost the signal. When unshielded twisted-pair is used, that distance shrinks to 100 m. The PMD layer specifies full-duplex connectors, optical transceivers, and bypass switches (optional).
- *PHY:* Physical layer (or the signal layer) makes the logical connection between the data link layer and the PMD. It handles encoding and decoding of information using a scheme called 4B/5B encoding. PHY specifies a set of line states that performs a handshake between PHYs in adjacent stations and uses a distributed clocking scheme to maintain ring synchronization at the physical level.
- *MAC:* Media access control (or the access layer) connects PHY to higher OSI layers. It performs functions similar to those of the IEEE 802 MAC layers. Its function is to schedule and transfer data on and off the ring. It also handles functions such as ring initialization, ring recovery, station address recognition, error control techniques, and the generation and verification of frame check sequence (FCS). It defines ring timers, frame formats, and the FDDI *timed-token protocol,* which is discussed in the following subsections. It also communicates with higher-layer protocols such as TCP/IP, IPX, and SNA.
- *SMT:* Station management (or the network management layer) controls the other three layers (PMD, PHY, and MAC) and ensures proper operation of the station. It handles such functions as initial FDDI ring initialization, station insertion and removal, ring stability, activation, connection management, address administration, scheduling policies, statistics collection, bandwidth allocation, performance and reliability monitoring, bit error monitoring, fault detection and isolation, and ring reconfiguration.

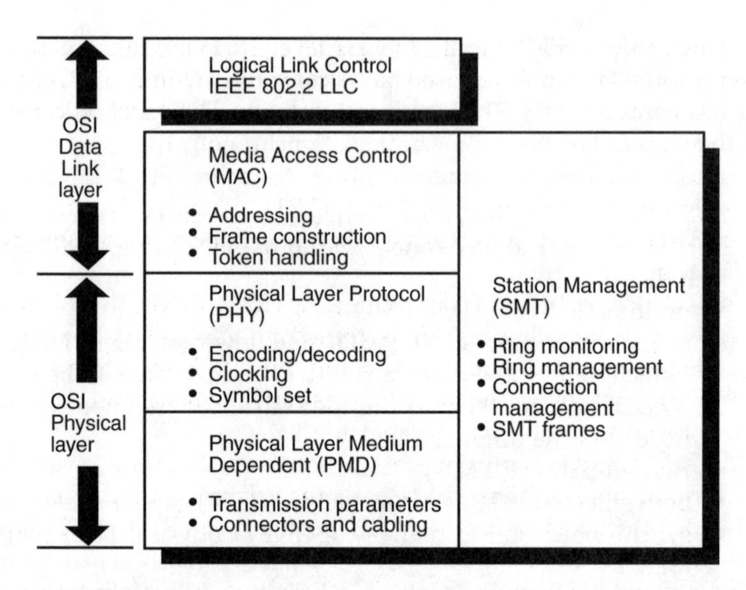

Figure 4.4 Summary of the functions of the FDDI standards.

These functions can be grouped into three categories: connection management, ring management, and frame services. Figure 4.4 summarizes the key functions of the PMD, PHY, MAC, and SMT in the FDDI architecture. The SMT standard took a long time to be approved, and it was not approved until 1993. It has been described as the most sophisticated multivendor interoperability document ever created.

4.1.2 Access and priority mechanism

Like a token ring LAN, access to the network in FDDI is gained by capturing a circulating token, a short packet that gives the transmission permit. When a station has a packet for transmission, it waits for a token. On receiving a token, the station transmits its packet before inserting the token in the ring again. The station passes the token immediately after transmitting the message and *before* receiving an acknowledgment that the message has been received. (Thus, unlike token ring LANs, packets from several stations can share the ring simultaneously. Also, because of the high data rate, it would be grossly inefficient to require stations to wait for their message to return, as in regular token ring LANs. This slight change in the protocol avoids wasting bandwidth.) All stations circulate the transmitted packet along the ring until it reaches its source station where it is extracted from the ring.

During the network initialization process, all stations connected to the ring negotiate for a target token rotation time (TTRT). Each station is assigned a portion of the TTRT for the transmission of its synchronous packets, while the remaining bandwidth is used for asynchronous traffic. The value of TTRT is determined small enough to satisfy the real-time constraints of synchronous traffic. This TTRT value (typically 8 ms) is set equal to the value of the token-rotation time (TRT), which is used to monitor the amount of elapsed time between subsequent arrivals of tokens at a station. Each station keeps track of the time it was last visited by a token. Synchronous traffic may be transmitted unconditionally upon token capture. Asynchronous traffic, however, is controlled by the token rotation time (TRT), which is the time interval between two successive arrivals of a token at a station. When running, the TRT counts down to zero.

If a token arrives before TRT expires (i.e., TRT < TTRT), the token is said to be early, and the value of TRT is copied into the token holding timer (THT), which begins to count upward, and the TRT is reset to measure the next token rotation. When the token arrives early, asynchronous frames may now be transmitted until THT reaches the value of TTRT. In other words, when the token arrives early, the station is permitted to send asynchronous frames for an amount of time that equals the earliness of the token. On the other hand, if the token arrives after TRT expires (i.e., TRT > TTRT), it is said to be late, and the station is allowed to send only synchronous frames. A token will arrive, on average, every TTRT seconds although it may arrive early or late on any given rotation. The TTRT value is selected so that the average TRT ≤ TTRT and the maximum TRT ≤ 2 × TTRT. In case different asynchronous priorities are implemented, each priority level j will have its own threshold TRT Pri(j) ≤ TTRT, j = 1, …, 8. The timed-token protocol has two important properties: (1) the average value of TRT is at most TTRT, and (2) the maximum TRT is at most twice TTRT. FDDI can support synchronous traffic only because TRT is bounded.

The use of the TTR protocol allows stations to request guaranteed bandwidth and response time for synchronous frames. It also establishes a maximum ring utilization equal to

$$\rho_{max} = \frac{\text{TRT} - \tau}{\text{TRT}} \tag{4.1}$$

where τ is the physical ring latency, the propagation time around the ring or the time for a token to go around an idle ring. If there are N active stations,

$$\text{TRT} = \text{N} \times \text{THT} + \text{token time} \tag{4.2}$$

where THT is the token holding time, the time a station is allowed to hold the token and transmit its frame. The token holding time is the difference between TTRT and measured TRT, i.e.,

$$\text{THT} = \text{TTRT} - \text{TRT} \tag{4.3}$$

The efficiency η of the timed token access method is the ratio of bandwidth used to carry packets to channel capacity. It is given by

$$\eta = \frac{N(T-\tau)}{NT+\tau} \tag{4.4}$$

where N is the number of active stations and T = TTRT. The maximum access delay is

$$D_{max} = (N-1)T + 2\tau \tag{4.5}$$

D_{max} is linear with N, but η approaches 1 if N and T become large. Therefore, making T large only slightly increases the ability to carry data but significantly increases the maximum access delay.

4.1.3 *Applications of FDDI*

The high bandwidth of FDDI permits new applications that were not previously possible. There are several major areas of application for FDDI. The obvious applications of FDDI include the following:

- *Backbone network:* FDDI is suitable for backbone applications for at least two reasons. First, its 100 Mbps data rate allows it to support the aggregate bandwidth requirement of several LANs. Second, FDDI can span distances up to 2 km between stations. A single FDDI network can connect as few as two LANs and as many as several hundred LANs. This allows the creation of a backbone network to link multiple buildings in a metropolitan area.
- *Front end:* FDDI is more commonly used for front-end applications. It typically consists of workstations, file servers, and computer servers connected together through concentrators. This provides the advantages of a star topology and improves the reliability of the ring. High-speed data transfer, expert systems, 3D, and full color graphics applications are excellent FDDI application niches.
- *Mainframe connectivity:* FDDI was initially conceived for high-speed mainframe-to-mainframe and mainframe-to-peripheral connectivity. Although this back-end use of FDDI is very important, it has not received much attention as backbone and LAN interconnection applications. With the mainframe market growing every year, there will be many of them to connect.
- *Government operations:* The U.S. Department of Defense (DoD), U.S. Navy, and NASA are all interested in FDDI because of its fault-tolerant nature and for the fact that it is an ANSI standard. An area of importance to the military is vehicular and on-board applications for ships, submarines, and aircraft. FDDI has been slated to be used for NASA and space station applications. FDDI is suitable for airborne installations because of its small size, light weight, and immunity to electromagnetic interference (EMI).

Other typical FDDI application areas include bandwidth-intensive data communications, such as graphics and CAD/CAM, installation near electric power, installations in high-EMI environments, and areas affected by lightning. FDDI has been used as a back-end (I/O channel) interface, front-end high-performance networks, and backbone networks for LAN interconnection. Various features of FDDI make it attractive for these and other applications.

4.1.4 Enhanced FDDI

Although the original FDDI, described above, provides a bounded delay for synchronous services, the delay can vary. This type of variable delay makes FDDI unsuitable for many constant bit rate (CBR) applications that require a strict periodic access. FDDI was initially envisioned as a data-only LAN. The full integration of isochronous and bursty data traffic is obtained with the enhanced version of the protocol, known as FDDI-II. FDDI-II is described by the hybrid ring control (HRC) standard, which specifies an upward-compatible extension of FDDI. FDDI-II adds one document, HRC, to the existing four documents that specify the FDDI standard. FDDI-II builds on the original FDDI capabilities and supports integrated voice, video, and data capabilities, but it maintains the same transmission rate of 100 Mbps. FDDI-II therefore expands the range of applications of FDDI.

Although FDDI-II has been implemented by several vendors, its future looks bleak. It may not become a widely implemented standard because of emerging networks, such as SONET, SMDS, and ATM, which address circuit-switched or isochronous services far better. ATM is in fierce competition with FDDI-II because it holds a bright future and has a great deal of backing.

FDDI seems to be fading rapidly into networking history. Users are migrating from FDDI because of diminishing vendor support, causing FDDI sales to diminish as well. An alternative to FDDI has been proposed in light of these problems, but that will not address all issues facing FDDI. FDDI is losing ground with other alternatives such as asynchronous transfer mode (ATM) and Gigabit Ethernet. Ultimately, it is not complexity, price, or standard issues that could dictate the end of FDDI, but its multimedia shortcomings — its inability to provide guaranteed delivery for time-critical multimedia traffic like voice and video.

For a complete formal description of FDDI, the reader is referred to the standard documents.[13–17] Other issues on the technology, applications, and design of FDDI and FDDI-II are addressed elsewhere.[3–5,18]

4.2 SONET

The Synchronous Optical NETwork (SONET) is a fiber optics–based network for use by telephone companies. SONET provides an open standard optical interface for transmission at the broadband user network and between network nodes. It is intended to provide a common international rate structure and eliminate the different transmission schemes and rates of nations. SONET is an emerging technology that will definitely have an impact on MANs and WANs.

Three primary needs have driven the development of SONET. First, there was the critical need to move multiplexing standards beyond the DS3 (44.7 Mbps) level. Second, there was the need to enable cost-effective access to relatively small amounts of traffic with the gross payload of an optical transmission. Third, there were data communication needs such as LAN traffic, video, graphics, and multimedia. SONET provides the infrastructure over which BISDN services can be deployed.

SONET was originally conceived by R. J. Bohm and Y. C. Ching of Bellcore (Bell Communication Research). It was proposed by Bellcore as a standard optical interface to the ANSI T1 committee in 1984. It is now an ANSI standard that defines a high-speed digital hierarchy for optical fiber. An international version of SONET is the Synchronous Digital Hierarchy (SDH). The SONET specification was designed for the United States and Canada, whereas the SDH specification was for Europe, Asia, and other nations.

SONET is not a service but a transport interface that enables a communication network to carry various types of services. SONET is also not a communications network in the same sense as LAN, MAN, or WAN are, but rather an underlying distribution channel over which communications networks such as FDDI, IEEE DQDB (dual-queue dual bus), and SMDS (switched multimegabit data service) can all operate. SONET then provides the basis for the optical infrastructure that will be necessary to meet the communications needs of the decade and beyond.

4.2.1 Basic features

SONET is:[1,19]

- A series of ANSI standards for fiber-optic data transmission defined for a hierarchy of rates (51.84 Mbps to 2.488 Gbps) and formats
- A layer 1 (physical layer) technology whose purpose is to carry any payload across distances at relatively high speeds
- A scalable network
- A transport network that uses synchronous operation with powerful, yet simple, multiplexing
- A system that provides extensive operation, administration, and maintenance services

SONET has the following basic features:[20]

- SONET has the ability to accept input from diverse sources and multiplex them into SONET frames.
- Since any tributary can be inserted into or extracted from an aggregate signal stream, significant savings in transmission hardware are possible.

- SONET standard emphasizes the need for operations, administration, and maintenance (OAM) of an end-to-end system. OAM refers to monitoring, managing, and repairing of a SONET system.
- SONET systems are designed to operate with central wavelengths at 1310 nm with SMF fibers or at 1550 nm with DS-SMF fibers.
- It supports asynchronous interfaces to existing digital signals and provides a smooth migration to synchronous network.
- Above SONET, one can implement any data link technology, including ATM, frame relay, FDDI, and SMDS.

These features provide high quality, network survivability, and self-healing attributes in addition to traditional transmission capability.

The existing networks in North America have three major drawbacks.[21] First, a multivendor environment is not feasible for optical transmission because interconnection between different vendors is not possible at the optical interface. Second, the signals must be multiplexed/demultiplexed into asynchronous standard electrical interfaces for interconnection between different vendors, which requires many multiplexers/demultiplexers. Third, the multiplexing hierarchy of DSO (64 kpbs), DS1 (1.544 Mbps), DS2 (6.312 Mbps), and DS3 (44.736 Mbps) has been standardized, but multiplexing hierarchies beyond DS3 are generally proprietary. SONET solves these problems and allows the implementation of a reliable, cost-effective, high speed, and broadband communication network. Thus, the rationale behind SONET includes the following:[22]

- *Standardization:* SONET offers seamless interconnection among carriers without depending on their proprietary capabilities.
- *Bandwidth management:* Regardless of carriers, bandwidth can be managed to provide maximum control and support for bandwidth-on-demand services.
- *Universal connectivity:* SONET interconnects with a variety of current and emerging carrier services including FDDI, DQDB, SMDS, and ATM.
- *Transmission rate:* SONET offers 51.84 Mbps to 2.488 Gbps and can extend to much higher rates.

4.2.2 Architectural layers

The SONET standard was developed using a layered approach. The four optical interface layers for SONET are shown in Figure 4.5. They correspond to the physical layer of the OSI model. The layers, from the bottom up, are:

- *Photonic layer:* This layer is mainly responsible for converting electrical STS-N frames (to be discussed later) to optical OC3-N frames. The layer handles the transmission of bits across the optical fiber. Electro-optical devices communicate at this layer. Other issues such as optical

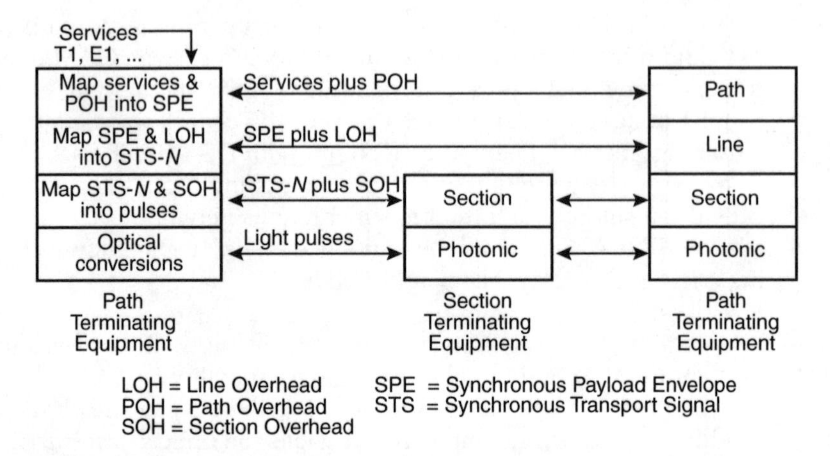

Figure 4.5 SONET interface layers.

pulse shape, clock recovery, jitter control, receiver and transmitter power levels, and operating wavelength are handled at this level. This layer is the only one without an associated overhead.

- *Section layer:* A section of the transmission facility includes termination points between either two repeaters or a repeater and a terminal network element (e.g., multiplexer). The section layer is responsible for transmitting STS-*N* frames across the transmission medium. Other functions include framing, scrambling, error monitoring, and adding the section-level overhead. Scrambling is necessary to avoid the occurrence of long ones or zeros. SONET optical interfaces, and all optical transmission systems that use binary coding, must scramble their transmission frame prior to transmission to ensure an adequate number of transitions (zeros to ones and ones to zeros) for line rate clock recovery at the receiver.
- *Line layer:* A line is the transmission medium that transports information from one network element to another. The line layer handles the transmission of path layer overhead and payload across the medium. It is also responsible for synchronization and multiplexing for the path layer.
- *Path layer:* A path corresponds to a logical connection between source and destination. The path layer essentially maps services into the format required by the line layer for transmission over the fiber. It deals with end-to-end transport of services between pieces of path-terminating equipment. It also provides for end-to-end communication via path overhead.

Besides the photonic layer, each layer has associated overhead. The relationship between section, line, and path is shown in Figure 4.6. The basic building block of a SONET is a *section* that consists of two network elements (NEs)

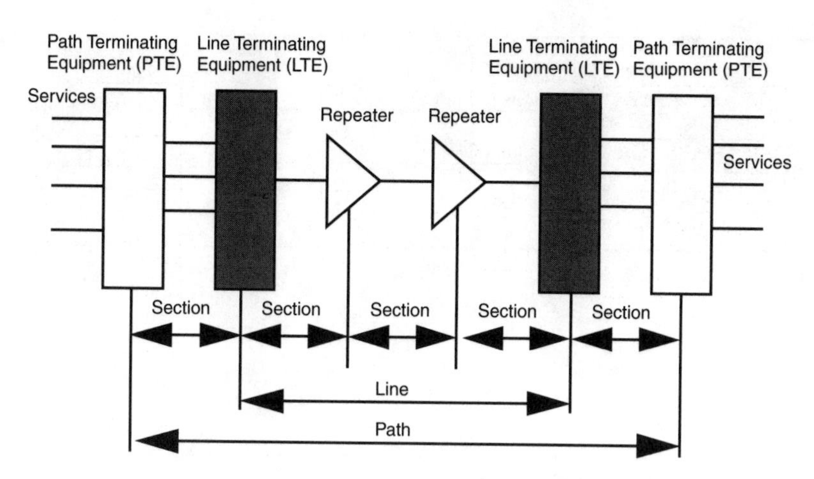

Figure 4.6 SONET configuration.

directly connected by optical fiber. Examples of NEs include synchronous multiplexers, add/drop multiplexers (ADMs), regenerators (or repeaters), and cross connects. A *line* consists of several sections. A *path* is an end-to-end connection between terminals.

4.2.3 Frame format

A synchronous signal is a set of bytes organized into a frame structure. In SONET, the basic unit of transport is the *synchronous transport signal level 1* (STS-1) frame, which has a bit rate of 51.84 Mbps and repeats every 125 μs, i.e., 8000 SONET frames are generated per second. The basic rate of 51.84 was selected partly because it could support interconnectivity between U.S. T-3 facilities running at 44.73 Mbps and European E-3 running at 34.36 Mbps.

The key features of the STS-1 framing include:

- Frame repetition of 8000 frames per second
- Frame duration of 125 μs
- Support for sub-STS-1 signals such as DS-1s

An STS-1 frame structure is generally presented in the characteristic matrix shape in which every element represents a byte. The frame structure is portrayed in Figure 4.7. It consists of 9 rows and 90 columns (9 × 90 bytes forming a total of 810 bytes per frame). Each byte in the SONET payload equates to a 64-kbps channel. Each rectangle in Figure 4.7 represents a 125-μs snapshot of a transmission signal. Transmission is serial (bit by bit) and occurs from left to right, row by row.

Flexibility of SONET is demonstrated by a feature called *virtual tributaries* (VTs). As shown in Figure 4.7(a), the STS-1 frame consists of virtual tributaries as needed by the application, such as data, voice, or video. A virtual

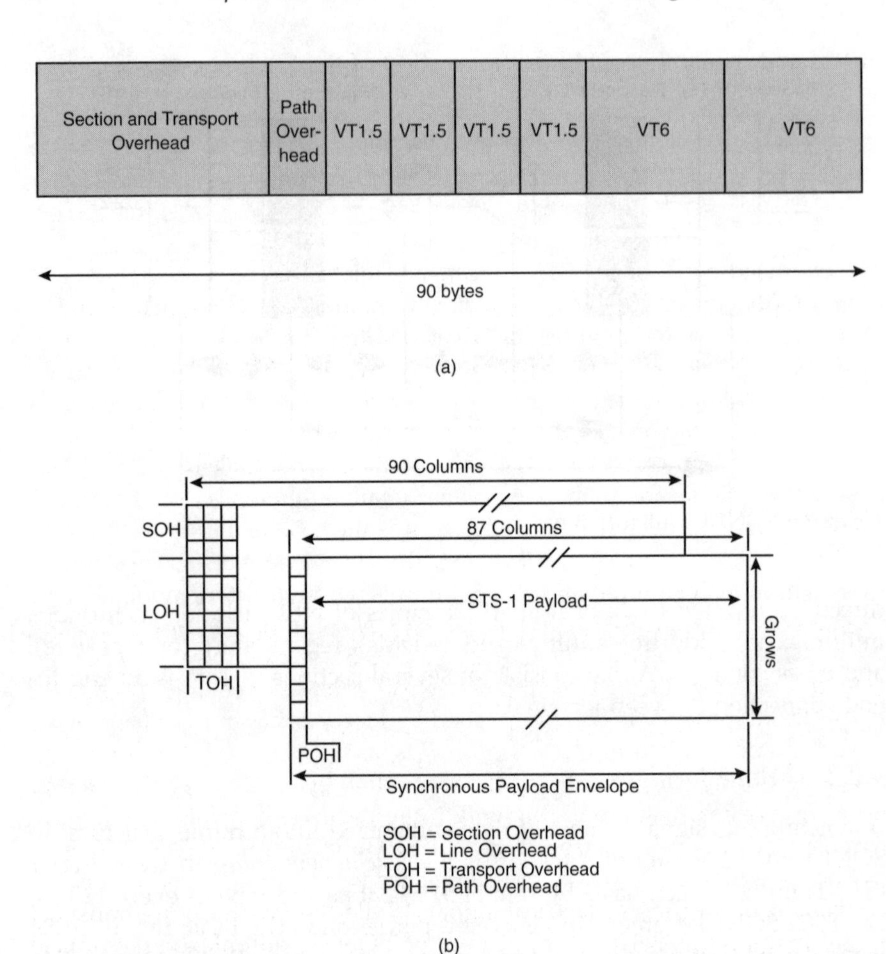

Figure 4.7 SONET STS-1 frame structure: (a) with the virtual tributaries, (b) without the virtual tributaries.

tributary is a structure defined in the SONET standard in order to maintain a consistent payload structure while providing for transport of a variety of existing lower rate services, such as 56 kbps, 64 kbps, DS1, DS2, T1, E1, or E4 signals. The lower speed signals are inserted and removed from the SONET frame using an Add/Drop (ADM) multiplexer. Table 4.1 shows different types

Table 4.1 Types of Virtual Tributaries

VT Type	(Rows, Columns) with STS SPE	Line Rate (Mbps)
VT1.5	(9,3)	1.728
VT2	(9,4)	2.304
VT3	(9,6)	3.456
VT6	(9,12)	6.912

of VTs and their rates. VT1.5 is the smallest of the VT types. It is a 27-byte structure, or 3 columns by 9 rows. Its bit rate is calculated as

$$27\frac{\text{bytes}}{\text{frame}}x8\frac{\text{bits}}{\text{byte}}x8000\frac{\text{frames}}{\text{sec}}=1.728\text{ Mbps} \tag{4.6}$$

The overhead bytes of the STS-1 frame are related to the section, line, and path components of a SONET network. As a matter of fact, SONET promotes the concept of layered maintenance by providing overhead channels at each of its path, line, and section layers. As shown in Figure 4.7(b), the STS-1 frame is divided into two areas: the *transport overhead* (TOH) and the *synchronous payload envelope* (SPE). The TOH carries overhead information, while the SPE carries the SONET payload, which may be ATM-based or STM-based. These fields are defined in greater detail in References 22 through 24.

In addition to the STS-1, which serves as the basic building block, there are higher rate SONET signals (STS-*N*). An STS-*N* channel rate is obtained by synchronously mutiplexing *N* STS-1 inputs (i.e., byte interleaving *N* STS-1 signals together). An STS-*N* frame has $90 \times N$ columns per row, including $4 \times N$ columns of interface overhead. An STS-3, for example, is formed by octet-interleaving three STS-1 frames. Byte or octet-interleaving is a SONET multiplexing process. It is accomplished by merging 1 byte from each input port with the bytes from other input ports, as opposed to bit interleaving. It is a procedure for interleaving the individual bytes of a signal such that each component signal is visible within the combined signal stream. Standards already define rates from STS-1 (51.84 Mbps) to STS-48 (2.48832 Gbps), as shown in Table 4.2, where optical carrier level *N* (OC-*N*) is the optical equivalent of an STS-*N* electrical signal. While the synchronous transport signal (STS) is the electrical input into the SONET multiplexer, the optical carrier (OC) is the optical output of the multiplexer. Other than scrambling, the optical signal (OC-*N*) is derived by bit-for-bit conversion from the electrical signal (STS-*N*). Thus, the two rates, formats, and framings are identical. Note from Table 4.2 that increments are in multiples of three: OC-1, OC-3,

Table 4.2 SONET Signal Hierarchy

Electrical Signal	Optical Signal	Data Rate (Mbps)
STS-1	OC-1	51.84
STS-3	OC-3	155.52
STS-9	OC-9	466.56
STS-12	OC-12	622.08
STS-18	OC-18	933.12
STS-24	OC-24	1244.16
STS-36	OC-36	1866.24
STS-48	OC-48	2488.32

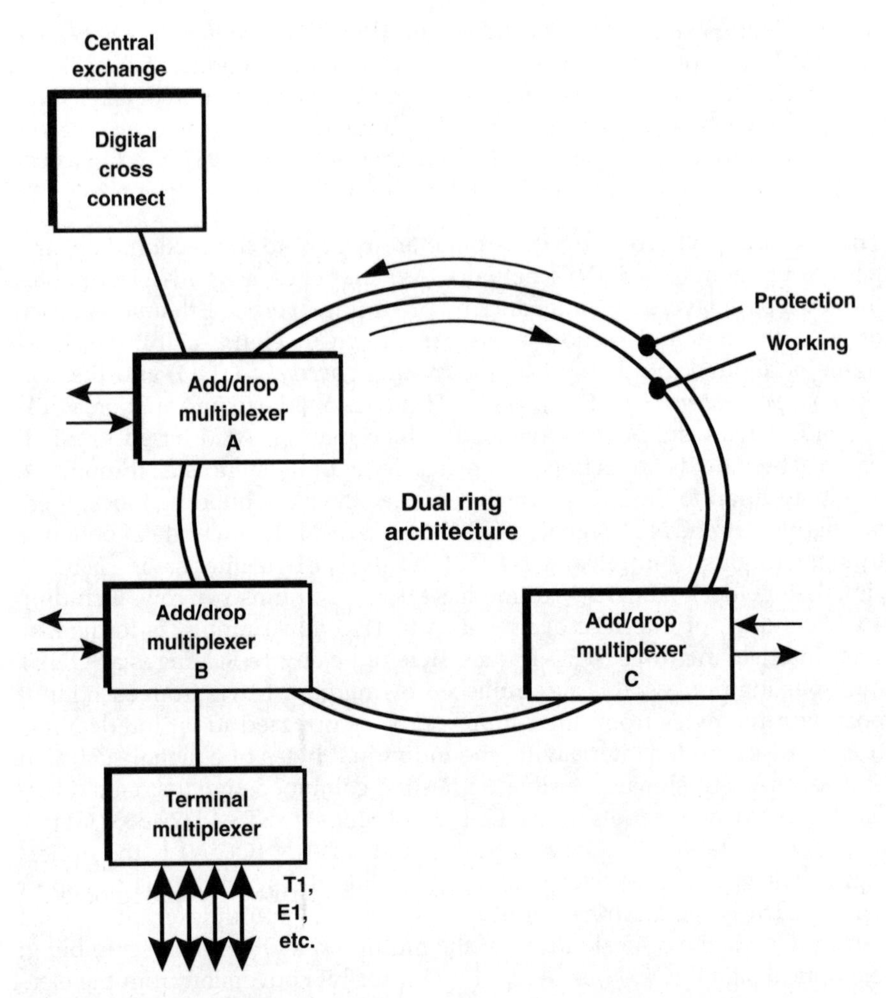

Figure 4.8 Typical SONET equipment.

OC-9, OC-12, OC-36, etc. The standard will eventually define rates up to STS-255 (13.2192 Gbps) or even OC-768.

Multiplexing allows several low-rate signals to be combined to form a higher-rate bit stream. Each low-rate signal is known as a *tributary* of the higher-rate multiplexed signal. Synchronous multiplexing of signals is the key feature of SONET/SDH. In synchronous multiplexing, the position of a tributary is fixed in the higher-order multiplexed signal.

4.2.4 *Equipment and topologies*

SONET network elements (NEs) refer to equipment that must be deployed to make SONET a reality. Figure 4.8 shows some of the major equipment used in SONET networks. Such equipment includes:

- *Fiber optical transmission systems* (FOTS): FOTS multiplex *N* STS-1 signals to form an STS-*N*, which is converted to an optical OC-*N* signal.
- *Add/drop multiplexers* (ADMs): ADMs can be in any part of the SONET network. They can drop any STS frame from within an optical carrier and route the STS frame onto another optical carrier or out to a non-SONET device. They can also be used in a central office to plug in, or add, circuits into a SONET facility. They are often placed in series with a SONET route and allow DS1 or other signals to be dropped or added to an STS-1 signal. ADMs allow access to SONET networks without having fully to demultiplex the SONET signal.
- *Terminal multiplexer (TM):* The TM is the end point on a SONET network. It is a network element that terminates several DS1 signals and assemble them into STS-1 payloads. It differs from the ADM in location and function, but they are fundamentally the same. TMs are typically equipped with an integrated electrical-to-optical converter.
- *Digital cross-connect system (DCS):* A cross-connect accepts various OC-*N* rates, accesses the STS-1 signals, and switches at this level. It is best used as a SONET hub, where it can be used for grooming STS-1s. A major difference between a cross-connect and an ADM is that a cross-connect may be used to interconnect a much larger number of STS-1. In addition to allowing switching and circuit grooming, DCS will improve the capabilities of network management, customer control, and network restoration.

SONET provides mechanisms for setting up a linear or ring topology, and the SONET equipment vendors provide software to configure the topology for addressing, cross-connect, and pass-through operations. It can transport a variety of traffic using point-to-point and ring-based topologies. A typical point-to-point (bus) topology is shown in Figure 4.9. This application is the simplest of SONET. The add-drop module (ADM) is a synchronous multiplexer that adds or drops a signal in a multiplexed stream. The cross-connect (an ATM switch) switches virtual channels.

A typical dual-ring topology is illustrated in Figure 4.10. Two fibers are used; one is a working fiber, while the other is a protection fiber. If a failure occurs, SONET nodes reverse the direction of traffic using the the other fiber, as in FDDI. The ADM can be reconfigured to ensure continuous operations in case of a ring failure, such as one caused by loss of signal (LOS), loss of frame (LOF), or loss of pointer (LOP). Thus, rings provide a higher degree of reliability. The dual-ring topology is recommended when network survivability is needed. While there are several advantages to the ring topology, it is more difficult to manage from the protocol's perspective.

4.2.5 Deployment and applications

The telephone network went through its first evolution when the analog equipment was replaced, one at a time, by digital counterparts. The deployment

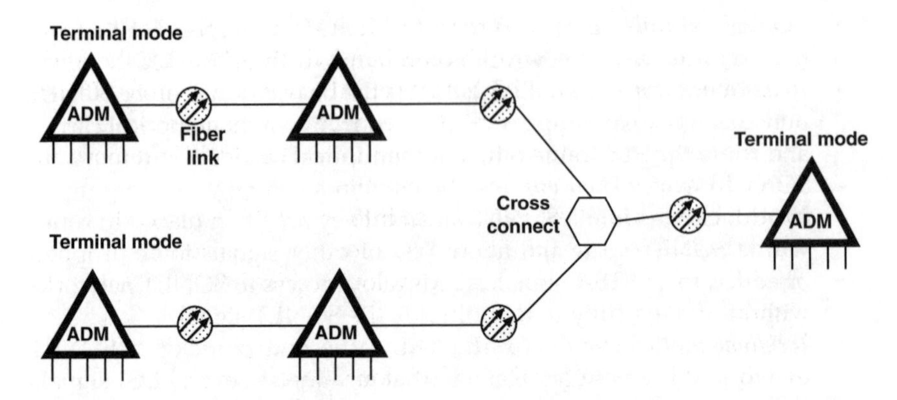

Figure 4.9 Typical SONET point-to-point topology.

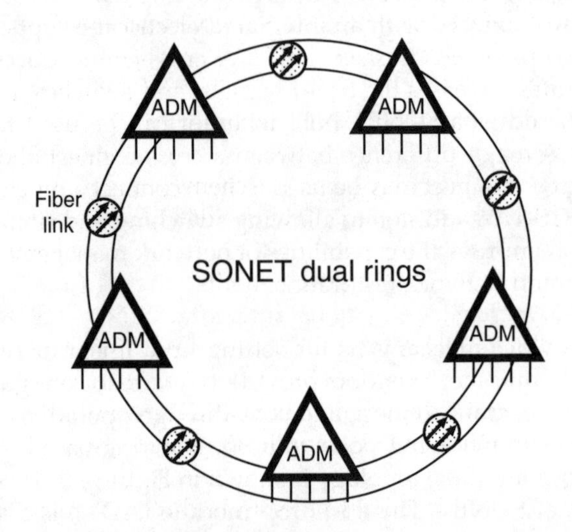

Figure 4.10 Typical SONET dual-ring topology.

of SONET equipment constitutes its second evolution: from an asynchronous digital hierarchy to a synchronous digital hierarchy.

SONET is fast becoming the *de facto* standard for high-speed broadband communications used by corporations, phone companies, universities, and others. This situation is because of SONET's advantages over current telecommunication technology. These advantages include:[25]

- It is an integrated network on which all types of traffic can be transported.
- It is a synchronous network, and hence it allows single-stage multiplexing and demultiplexing.
- It provides a long-term solution for an optical mid-span meet between vendors, i.e., it allows equipment from different vendors to communicate with one another.

- SONET is built on optical fiber technology that provides superior performance and almost unlimited bandwidth. Although the initial investment costs remain high, the optical hardware is more efficient and cost-effective in terms of maintenance and management.
- The management features that are built into the SONET frame structure will help in maintaining the circuits.
- Protection and fault tolerance built into the SONET network design will translate into additional revenue for service providers.
- The use of synchronous ring topologies provides high network resilience.
- It is a universal standard and nonproprietary.
- SONET employs a digital transmission scheme.
- SONET is backward and forward compatible; it can handle both current and future services.
- It supports FDDI, DQDB, SMDS, ATM, and other high bandwidth services.

Some of the drawbacks of SONET include:

- It requires strict synchronization schemes.
- It lacks a control system and bandwidth management.

The deployment of SONET will offer benefits ranging from improved network control and efficiencies to global interconnection.[26,27] SONET will find applications in BISDN, FDDI, and SMDS.

What applications will drive SONET network infrastructure? Four key factors that will affect the deployment of SONET are flexibility, survivability, reliability, and economics. SONET applications include:[28]

- *ATM:* ATM is an international standard for cell relay technology. It is the multiplexing and switching mechanism for a BISDN network. SONET rings are used to interconnect ATM switches. Since ATM cells are asynchronous, they float within the payload and SONET pointers locate their positions. For certain, SONET will dominate the transport and switching facilities of public networks and will support a variety of services.
- *LAN interconnection:* WANs often provide DS1 (T1) or DS3 (T3) service. Both of these services have their corresponding signaling rates which easily map into the SONET SPE. SONET technology can also be used in a MAN environment that connects many LANs together. LAN interconnection is another good application for the bandwidth that SONET provides. Without competing with the enormous speeds of Fast Ethernet and Gigabit Ethernet, the delivery of LAN interconnection on SONET has much to offer a customer.
- *Video and graphics:* Many insurance companies, financial institutions, medical x-ray imaging, and the publishing industry are large users

and generators of graphic and video files. These large files are stretching PC hard drives and LANs to their limits. SONET, with its enormous bandwidth, offers the best hope for running these multimedia applications on a WAN. SONET will also allow high definition television (HDTV) to be experienced in the home. Transmission of HDTV via SONET will require compression to OC-3 rates of 155 Mbps.

- *Cinema of the future:* Cinema is one the exciting applications for SONET. For example, ATM switches can be combined with SONET to deliver major motion pictures to movie screens. The ATM multicasting switches can distribute video streams to a vast number of theaters simultaneously, possibly across the nation.

An important application of SONET is in improving network survivability. SONET self-healing rings have emerged as a widely accepted approach to this application.[29] Studies have shown great advantages of network architectures based on rings rather than on traditional hub structures. In the self-healing architecture, traffic between two points is carried in an add/drop network that is looped back on itself, forming a "ring." A ring protects itself against a fiber cut or a node failure by providing an alternate path around the failure, as illustrated in Figure 4.11. With the rapid decrease in the cost of transmission relative to switching, it is feasible to consolidate local exchange networks in SONET self-healing rings. This consolidation has the advantage of reducing the amount of transmission and switching equipment while offering significant improvements in network performance, reliability, and survivability.

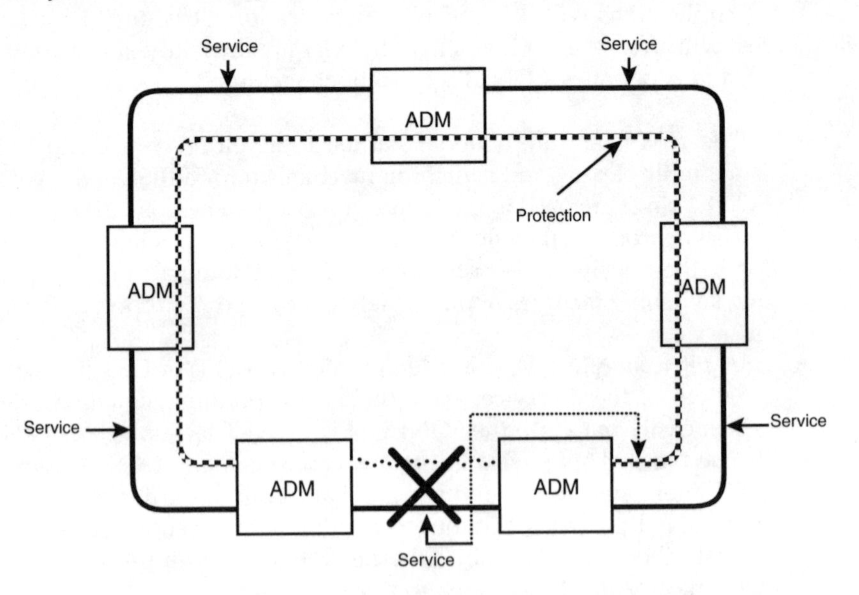

Figure 4.11 A cable cut or node failure in SONET self-healing rings.

Another area of tremendous growth and a potential SONET application is the Internet and World Wide Web, with its millions of users and with more Web pages appearing daily. The explosive growth in Internet traffic has created the need to transport IP on high-speed links. IP backbone providers are seeking cost-effective solution for providing interconnections between gigarouters. IP-over-SONET technology is a leading candidate to provide the solution to this problem. Some IP evangelists have claimed that packet-over-SONET (POS) is next in line to make ATM obsolete. Although POS is a viable technology, its purpose is too narrow — sending data at high rates.

WDM (wavelength division multiplexing) is an emerging technology developed to work within SONET systems and to increase dramatically the amount of bandwidth available on a fiber. It allows the simultaneous transmission of light over a fiber cable at different wavelengths. This increased capacity comes without the need to install more fiber in the SONET network.

However, there are a number of limitations inherent to ATM/SONET architecture:

- It is optimized for time-division-multiplexing (TDM) voice, not data. Its fixed bandwidth channels have difficulty handling bursty data traffic.
- It requires inefficient and expensive optical-to-electrical-to-optical conversion at each network node.
- It is hard to provision since each network element in the ATM path must be provisioned for each service.
- It requires installation of all nodes up front because each node is a generator.
- It does not scale well due to its connection-oriented virtual circuits.

A facilitator in the deployment of SONET is the SONET Interoperability Forum (SIF), an open industry group sponsored by the Alliance for Telecommunications Industry Solutions (ATIS). SIF is a collection of equipment vendors who seek to work together to overcome the technical hurdles to rapid deployment of SONET technology. It was established in 1991 to resolve SONET interoperability and implementation issues.

In North America, SONET will play a significant role, both economically and technologically, in the evolution strategy toward broadband. The FDDI, DQDB, SMDS, and ATM can all operate over SONET. Since these networks will be used for LAN interconnection, the potential importance of SONET in the corporate LAN is obvious. SDH will play similar roles in Europe, Asia, and elsewhere. Although differences exist in the basic frame format, SONET and SDH are the same beyond the STS-3 signal level, as evident from Table 4.3. Survivability and high bandwidth are SONET's major contributions to transmission technology. More about SONET/SDH can be found in References 1, 30, and 31.

Table 4.3 SDH/SONET Transmission Rates

SDH Signal	SONET Signal	Transmission Rate (Mbps)
—	STS-1	51.84
STM-1	STS-3	155.52
STM-4	STS-12	622.08
	STS-24	1244.16
STM-16	STS-48	2488.32

4.3 Fiber channel

Fiber channel is part of the recent revolution in computer communications caused by incorporating serial transmission line design techniques and technologies into applications in computer architecture that have conventionally used bus-based types of data transport. It is a highly reliable, gigabit interconnect technology that serves as a means of quickly transferring data between supercomputers, mainframes, workstations, desktop computers, storage devices, and other peripherals. It allows concurrent communications among those computer devices using SCSI (small computer systems interface) and IP protocols. The standard allows payload bit rates of 200, 400, or 800 Mbps.

There are basically two types of data communication between processors and peripherals: networks and channels. A network consists of distributed nodes and supports interaction among these nodes. A channel provides a direct or switched connection between devices over very short distances. While a channel is hardware-intensive and transports data at high speed with low overhead, a network is software-intensive, with high overhead, and slower speed than a channel. A fiber channel combines the best features of the two methods of communication into an I/O interface that meets the needs of both the network users and channel users. It has the flexibility and interconnectivity characterizing a protocol-based network while maintaining the simplicity and speed associated with channel connections. Although it is called fiber channel, both optical and electrical media are supported. (Thus *fiber* will be used as a generic term to mean an optical or a copper cable.) Also, its architecture does not represent either a channel or a network.

4.3.1 Basic features

Fiber channel is a network technology for linking servers to storage systems that move data more efficiently than existing network pipes. It is well suited for companies that want to build their own SANs (storage area networks).

Development started in 1984, and ANSI standard approval came in 1994, fiber channel was designed to remove the barriers of performance existing in legacy LANs and channels. The performance-enhancing features of fiber channel for networking include:[32]

- *Unification of technologies:* Fiber channel is geared toward unifying LAN and channel communications. It achieves this by defining an architecture with enough flexibility and performance to fulfill both sets of requirements.
- *Bandwidth:* Fiber channel provides more than 100 Mbps for I/O and communications on current architectures.
- *Inexpensive implementation:* By limiting low-frequency components, fiber channel allows design of receivers using inexpensive CMOS VLSI technology.
- *Flexible topology:* Physical topologies are defined for point-to-point links, packet-switched network topologies, and shared-media loop topologies.
- *Flexible service:* Fiber channel offers classes of services including dedicated bandwidth between port pairs, multiplexed transmission, and multiplexed datagram transmission.
- *Standard protocol mappings:* Fiber channel can operate as a data transport mechanism for multiple protocols with defined mappings for IP, HIPPI (high-performance parallel interface), IPI (intelligent peripheral interface), SCSI (small computer systems interface), ESCON (enterprise serial connection, developed by IBM), and ATM AAL5.
- Support for distances up to 10 km.
- Complete support for traditional network self discovery.
- Availability of both real and virtual circuits.
- Use of 8B/10B encoding, which allows design of gigabit receivers using inexpensive CMOS VLSI technology.
- Low overhead providing efficient usage of the bandwidth.
- Full support for time-synchronous applications, such as video, using fractional bandwidth virtual circuits.
- Efficient, high-bandwidth, low-latency transfers using variable length frames (0–2 kB).
- Two data paths to provide redundant connections; full duplex links with each having two fibers.
- A serial optical and coaxial system operating from 100 to 800 Mbps.

Fiber channel is a network that performs channel operations or a channel with network extensions. It uses an asynchronous mode of transmission permitting 100% utilization of the available bandwidth. It is scalable over a broad range of transfer rates, media types, and protocols.

4.3.2 Architecture

The fiber channel standard (FCS) is made simple in order to minimize implementation cost and enhance throughput. It consists of five levels as shown in Figure 4.12. The idea of levels is similar to that of layers in networks. Levels allow elements in each level to change (with technology or other needs) without affecting adjacent levels.

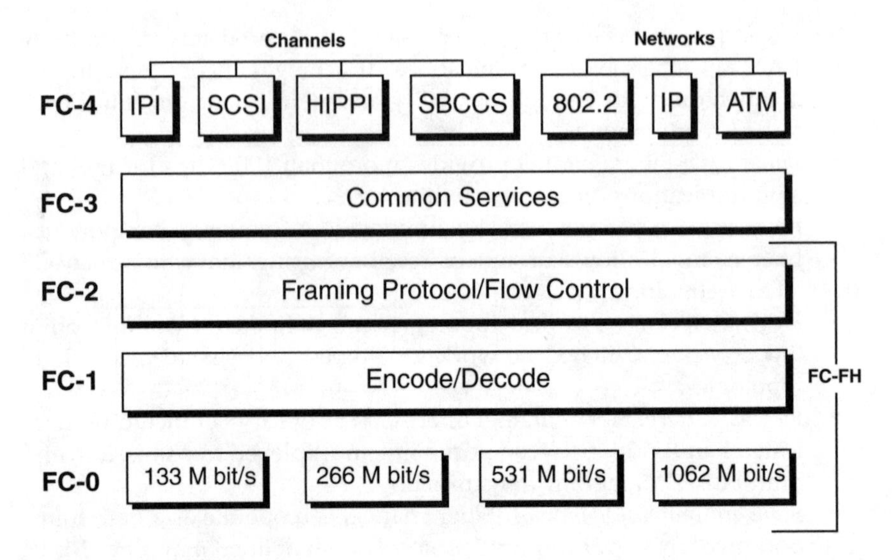

Figure 4.12 Structure of fiber channel standards.

The FC-0 level defines the physical portion of the fiber channel including the transmission media, connectors, transmitters and receivers and their interfaces, and optical and electrical parameters for a variety of data rates. The media can be optical cable with laser or LED transmitters for long distance transmission; copper coaxial cable for high speeds over short distances; or shielded twisted pair for low speed over short distances. The data transfer rates are defined as 100, 200, 400, and 800 Mbps. FC-0 operates with a BER of less than 10^{-12}. The FC-0 specification provides for a large variety of distances, cable plants, and technologies. Each variant is described with one code from each of the four characteristics, as shown in Figure 4.13. For

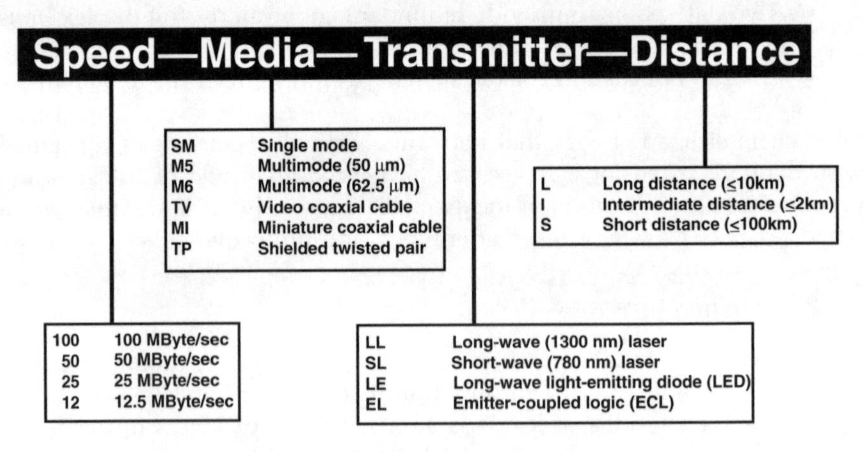

Figure 4.13 Fiber channel FC-0 nomenclature.

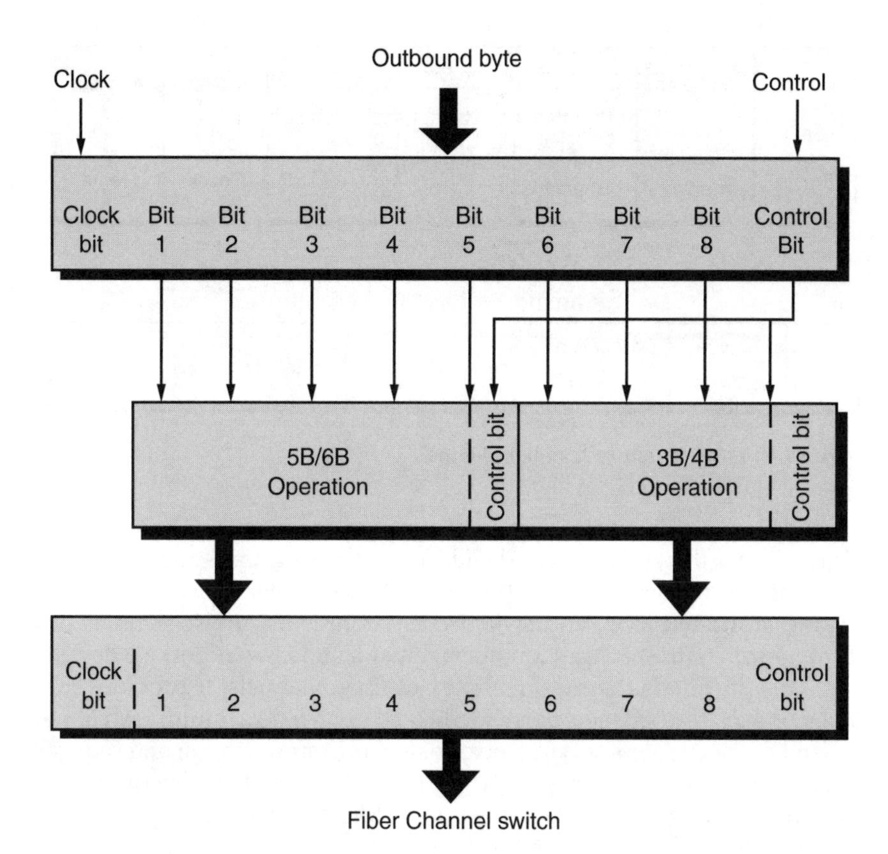

Figure 4.14 Fiber channel 8B/10B encoding scheme.

example, code 100-SM-LL-L indicates speed is 100 Mbps, medium is single-mode fiber, transmitter is long-wave laser, and distance is long.

FC-1 level describes the transmission protocol, which includes byte serial encoding/decoding, error control, and synchronization. This layer is responsible for establishing rules for transmitting information on the network. It handles an 8B/10B encoding scheme shown in Figure 4.14. An 8B/10B scheme is used just as the 4B/5B technique used in FDDI but it is more powerful than 4B/5B. The information transmitted over a fiber is encoded 8 bits at a time into a 10-bit transmission character. The information received is recovered 10 bits at a time, and those transmitted characters are decoded into one of the 256 8-bit combinations. The rationale for the transmission code is to facilitate error correction and bit synchronization. In other words, the scheme allows the protocol to synchronize data transmission and provides a retransmission mechanism when errors are detected. In addition, the 8B/10B encoding/decoding scheme is simple to implement and provides DC balance for the transmitted bit stream. The 8B/10B scheme was developed and patented by IBM for use in its ESCON interconnect system.[33,34]

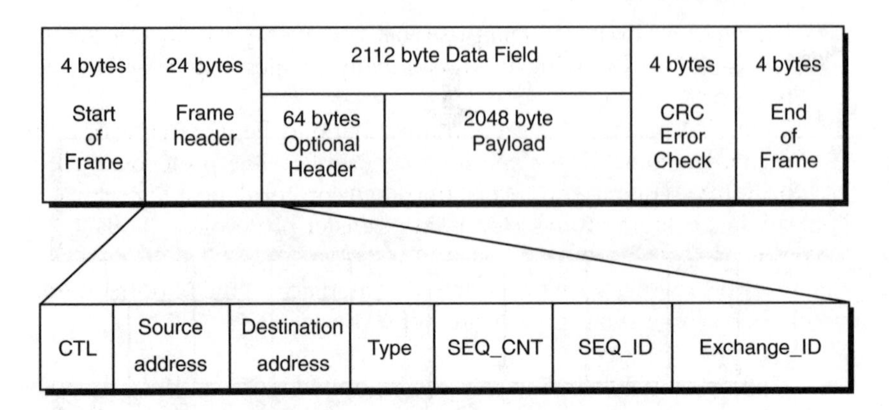

Figure 4.15 Frame structure for fiber channel.

FC-2 level is the framing and signaling protocol layer. (The frames are of variable length, with fixed overhead.) The level is the transport mechanism specifying the rules required to transfer data. The layer provides for point-to-point, arbitrated loop, and switched environments. It also handles flow control, error control, congestion management, and class of service designations. This portion is the most complex of fiber channels. It provides three classes of services between ports (dedicated connection, multiplex connection, and connectionless service), error detection, packetization and sequencing, segmentation and reassembly, and flow control between ports. The following building blocks are defined in FCS to aid in the transport of data across the link:[35–37]

- *Ordered sets:* These are four-byte transmission words containing data and special characters which have special meaning. The frame delimiters, the start-of-frame (SOF) and end-of-frame (EOF) ordered sets, immediately precede or follow the contents of a frame. An idle primitive indicates a port is ready for frame transmission and reception. The primitive Receiver Ready indicates that the buffer is available.
- *Frame:* A frame contains the information (payload), the SOF and EOF delimiters, the address of the source and destination ports and link control information. The frame structure is shown in Figure 4.15. There are two types of frames: data frames and link control frames. It is the responsibility of the FC-2 level to break the data into frame size and reassemble the frames.
- *Sequence*: A sequence is formed by a series of one or more frames transmitted unidirectionally from one port (the Sequence Initiator) to another (the Sequence Recipient). Each sequence is uniquely specified by the Sequence Identifier (SEQ_ID), as in Figure 4.15. The SEQ_CNT in Figure 4.15 is a counter field that uniquely identifies frames within a sequence; the first frame starts with 0 and SEQ-CNT is incremented for subsequent frames.

- *Exchange*: An exchange comprises a series of one or more nonconcurrent sequences flowing unidirectionally or bidirectionally between two ports.
- *Protocol:* A protocol is a set of frames that may be sent in one or more exchanges to achieve a specific purpose. Possible protocols include primitive sequence protocols, port login protocol, port logout protocol, fabric login protocol, and data transfer protocol.

FC-3 level provides a set of services common across the N ports of fiber channel node. Three such services are defined as:

- Multicasting, which delivers a single transmission to multiple destination ports.
- Stripping, which uses multiple ports in parallel to transmit a single information across multiple links simultaneously.
- Hunt group, which is the ability for more than one port to respond to the same alias address. Hunt groups are like a telephone hunt group where an incoming call is answered by any one of the phones; frames addressed to a hunt group are delivered to any available port within the hunt group.

FC-4 level defines the mapping of various channels and application interfaces that can execute over fiber channel. Fiber channel supports the following channel and network protocols:

- Intelligent peripheral interface (IPI)
- Small computer system interface (SCSI)
- High performance parallel interface (HIPPI)
- Single byte common code set mapping (SBCCS)
- IEEE 802.2 local area network (LAN)
- Internet protocol (IP)
- ATM (asynchronous transfer mode) adaptation layer (AAL5)
- Link encapsulation

The lower three levels (FC-0, FC-1, and FC-2) together form the fiber channel physical layer FC-PH, while the upper two layers are for common services and protocol interface. Several standards have been produced for level FC-4 to map a variety of channel and network protocols to lower levels. Fiber channel can support data, video, and voice at the same time, although each type of information has different characteristics and makes demands on the sender and receiver of information.

4.3.3 Topologies

Fiber channel devices are known as nodes. Each node has at least one port to access other ports in the other nodes. A node may be a computer, workstation,

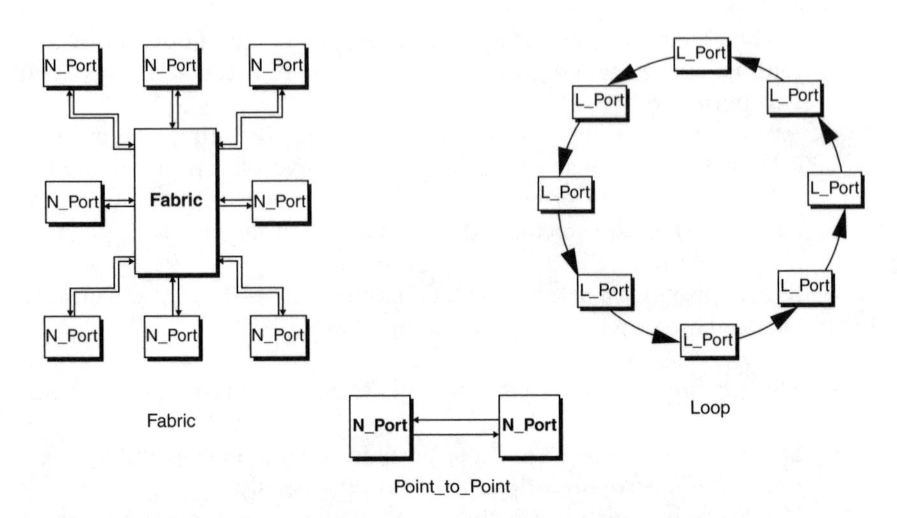

Figure 4.16 Fiber channel topologies.

a disk drive, TV camera, or a display. A port is the hardware entity within a node that permits data communication over a link. Each port uses a pair of fibers; one carries information into the port while the other carries information out. This pair of fibers is known as a link. An N_Port is a port at the end of a link and "N" stands for node. An F_Port is an access point of the fabric and "F" stands for fabric.

In fiber channel, the switch connecting device is called a *fabric*. In other words, a fabric is an entity that allows interconnection of any N_ports to it, that is, it provides switching and routing. The fabric is self-sustaining; nodes do not need station management. This greatly simplifies implementation. The link is the two unidirectional fibers transmitting to opposite directions with their associated transmitter and receiver.

A topology is a set of hardware components, such as media, connector, and transceivers, connecting two or more N_ports together. As shown in Figure 4.16, there are three topologies defined for fiber channel: (1) a point-to-point link between two ports, called *N_Port*, (2) a network of N_Ports, each connected to an F_Port into a switching network or fabric, and (3) a ring topology known as an *arbitrated loop*, which allows multiple N_Port interconnections without switching elements. The fabric topology is the most general topology. The arbitrated loop topology is a simple, low-cost scheme for connecting up to 126 nodes; the ports in the loop must combine the functions of the N_Ports and F_Ports, making them NL_Ports. Each topology has its own standard dealing with issues unique to the topology.

Fiber channel enjoys the following advantages:

- Multiple topologies meet application requirements.
- Multiple protocols can take advantage of the high-performance, reliable technology.
- High efficiency due to very little overhead.

- It is scalable — from point-to-point links to integrated enterprises with hundreds of servers.
- It offers gigabit bandwidth now.

Network applications of fiber channel include:

- High-performance CAD/CAE network
- Quick-response network for imaging applications
- Nonstop corporate backbone
- Movie animation
- Storage backup and recovery systems

4.4 Broadcast-and-select WDM networks

Beyond the architectures provided by FDDI, SONET, and fiber channel, we now consider all-optical WDM-based networks. These networks are of two types: (1) broadcast-and-select, and (2) wavelength-routed networks. Broadcast-and-select networks generally employ passive optical buses, stars, or wavelength routers for LAN applications. Wavelength-routing networks consist of nodes interconnected by point-to-point links and employ active optical components for WAN applications. Broadcast-and-select networks are discussed in this section and wavelength-routed networks are discussed in Section 4.5.

4.4.1 Topologies

There are two types of topologies for broadcast-and-select networks: star and bus. In the star topology, shown in Figure 4.17(a), nodes transmit on different wavelengths and are connected to a passive star via two-way fibers. The signals are optically combined by the star coupler and broadcast on the receiver fibers to all the nodes. Each node employs an optical tunable filter to select the wavelength addressed to it. This kind of network can still function if one node is down. The passive star network can also support multicasting because any number of nodes may tune to the same wavelength. The passive-star WDM network has the following three advantages.

- The switching element is centralized but it relegates the switching functions to the end nodes.
- It supports multicast or broadcast services.
- The passive star can be much cheaper because it involves very little electronics.

In the bus topology, shown in Figure 4.17(b), the bus is a dual-bus topology with two unidirectional folded buses carrying information in opposite directions. Nodes transmit into the bus through a coupler and receive from the bus through another coupler.

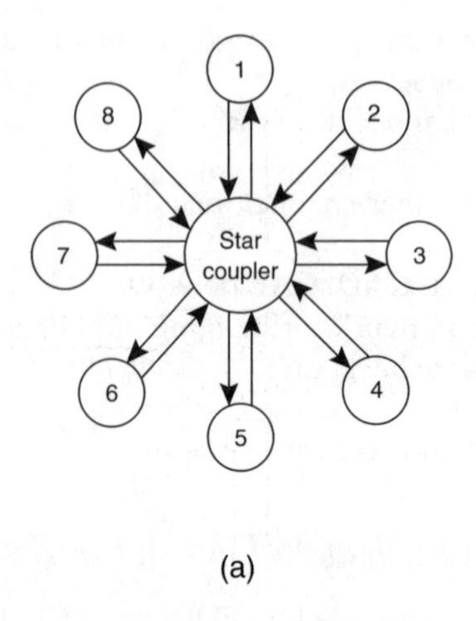

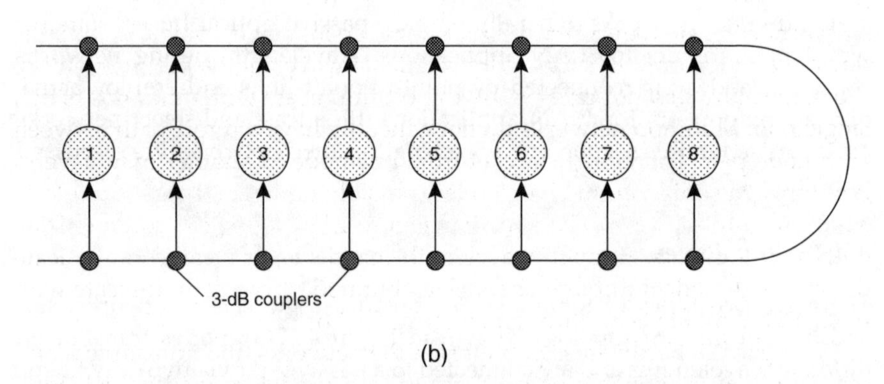

Figure 4.17 Broadcast-and-select topologies: (a) star, (b) bus.

Although there are few cases in which the bus topology outperforms the star, the star topology is a better choice: it can support a larger number of users than can a bus topology. This is because power loss and tapping loss in buses limit the number of users that can be connected without the addition of optical amplifiers. With the recent development of the erbium-doped broadband amplifier (EDFA), interest in the bus topology has been revitalized.

Communication between transmitters and receivers may follow one of two methods, leading to two types of broadcast-and-select networks: single-hop and multihop. A single-hop network is one in which information transmitted optically reaches its destination without being converted to an electrical form at intermediate points. It assumes that a tunable device can access all the available wavelengths, implying that any node can reach any other node. Networks in Figure 4.17 can be regarded as single-hop. In a multihop

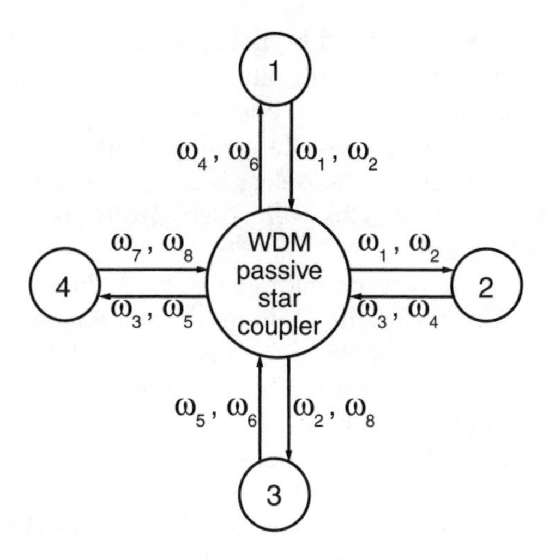

Figure 4.18 Example of typical multihop network.

network, intermediate electro-optical conversion takes place. A tunable device can tune only to some of the available wavelengths but not to all of them, meaning that not every node can communicate with other nodes in a single hop. Multihop networks do not generally have a direct path between each node pair. Therefore, a packet from one node to another may have to hop through some intermediate nodes, possibly none at all. An example of a typical multihop network is shown in Figure 4.18. Node 1 can communicate with node 2 directly through wavelength channel ω_1 or it can communicate directly with node 3 through channel ω_2. But node 1 can communicate with node 4 only by hopping through either node 2 or 3.

In regard to hardware capabilities of the networks, the transmitters and receivers may be tunable or fixed-tuned to a predetermined wavelength, leading to four possible structures:

- Fixed transmitters and fixed receivers.
- Fixed transmitters and tunable receivers.
- Tunable transmitters and fixed receivers.
- Tunable transmitters and tunable receivers.

If each node has a fixed transmitter and fixed receiver and all nodes are tuned to the same wavelength, it is a single-(wavelength) channel network. Single-channel networks include Ethernet, token ring, and FDDI.

4.4.2 Testbeds

To provide a picture of the current state of optical networking, recently developed broadcast-and-select testbeds are described below. Popular ones include LAMBDANET, Rainbow, and STARNET.[38–43]

LAMBDANET was developed by Bellcore. It employs a combination of time and wavelength division multiplexing. Each node has a fixed-tuned transmitter and an array of receivers, one of each wavelength. The transmitter uses time division multiplexing to send data to other nodes, while the receiver uses electronic circuits to select the traffic destined for it. Experiments report that 18 wavelengths were successfully transmitted at 2 Gbps over a distance of 57.5 km.

Rainbow-I network was developed as a MAN by IBM to support 32 wavelengths, 1 nm apart, with 32 nodes communicating with each other at 300 Mbps. It uses a star topology with each node having a fixed optical transmitter and a tunable receiver. The transmitter emits light at a wavelength different from that used by other transmitters in the network. A star coupler combines the traffic from different nodes. The network does circuit switching only. It is not suitable for packet-switched traffic because of the long request-response delays necessary before each packet transmission. A Rainbow-I network node consists of an IBM PS/2 with an adapter card as well as the hardware for a connection setup protocol. The connection setup protocol is a simple polling protocol. It uses both wavelength and time division multiplexing. Rainbow II is a follow-on to Rainbow I and uses the same hardware and MAC protocol as Rainbow I. It is an optical MAN that spans a distance of 10 to 20 km and supports 32 nodes, each operating at 1 Gbps. The nodes are attached to host computers via HIPPI.

STARNET is a WDM LAN developed at Stanford University. It is a high-speed optical packet-switched network based on passive star topology. Each node has a fixed-wavelength transmitter and two receivers — one main receiver that operates at 2.5 Gbps and an auxiliary receiver that operates at 125 Mbps, the same rate as FDDI. The network is intended for backbone applications and has been shown to be highly suitable for high-speed multimedia network applications. The first version of STARNET, known as STARNET-I, is based on coherent detection technology. The second version, STARNET-II, uses direct detection and subcarrier multiplexing to achieve the same objective. Direct detection does not require an additional local oscillator laser and is less sensitive to temperature fluctuations.

4.5 Wavelength-routed networks

Broadcast-and-select networks are not suitable for WANs for many reasons. First, as the number of nodes increases, more wavelengths are required. Second, since a broadcast-and-select network employs a passive star coupler, the network cannot interconnect a large number of nodes. Wavelength-routed networks are designed to overcome these limitations.

Wavelength-routed networks are considered to be potential candidates for the next generation of WANs. They are highly scalable because of wavelength reuse and suitable for WANs or backbone networks. A wavelength-routed network consists of wavelength routers and fiber links that interconnect them.

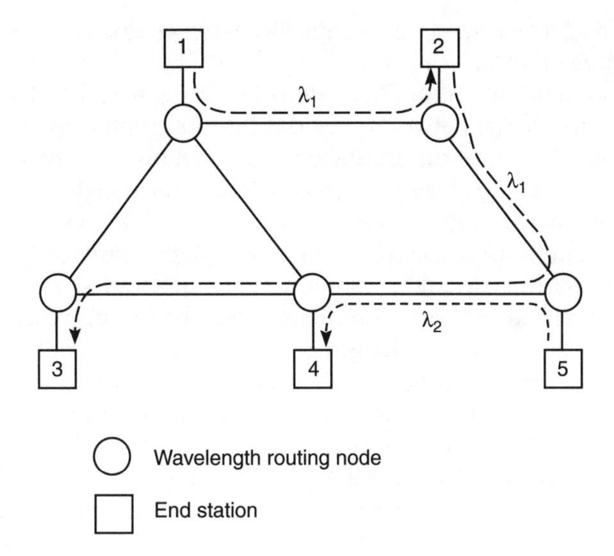

Figure 4.19 A typical wavelength-routing mesh network.

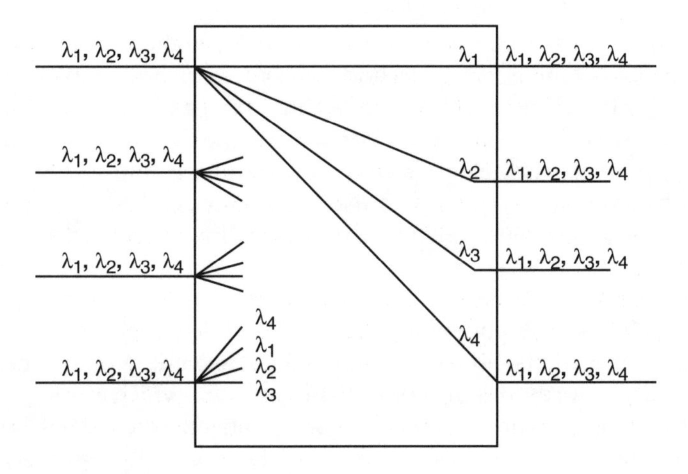

Figure 4.20 Wavelength router.

4.5.1 *Topologies*

The topology of a wavelength-routed network may be arbitrary as shown in Figure 4.19. The network consists of wavelength crossconnect (WXC) nodes with wavelength routing elements. As shown in Figure 4.20, the wavelength router is an optical switch capable of routing light signals from one input port to an output port. The nodes are interconnected by pairs of point-to-point fiber links with each link supporting a particular number of wavelengths (from 4 to 32 wavelengths). Each node is equipped with a set of transmitters and receivers, both of which may be wavelength tunable. A transmitter at a node sends data into the network and a receiver receives data from the network.

The basic mechanism of communication in a wavelength-routed network is a *lightpath*. The network provides lightpaths between node pairs. A lightpath is an all-optical communication channel between two nodes in the network, and it can span more than one link. It is a high-bandwidth pipe made possible by assigning a wavelength on each link on the path between two nodes. Lightpaths can also be regarded as performing end-to-end direct optical conversions without any intermediate electronics. The same wavelength can be reused to carry multiple connections through these devices. A lightpath uses the same wavelength on every link in its path. This requirement is the *wavelength continuity* constraint of the lightpath. Two lightpaths that share a link must use different wavelengths; the same wavelength cannot be allocated to two lightpaths on a link. This constraint prevents lightpaths from interfering with each other. For example, in Figure 4.19, lightpaths are established between nodes 1 and 2 on wavelength λ_1, between 2 and 3 on wavelength λ_1, and between 4 and 5 on wavelength λ_1. Note the wavelength reuse of λ_1.

In a network with N nodes, one would ideally establish lightpaths between all the $N(N-1)$ pairs. However, such a setup is not possible for two reasons. First, the number of wavelengths available imposes a limit on how many lightpaths can be set up. Second, each network node can be the source and sink of only a limited number of lightpaths. The intermediate nodes in the fiber path use their active switches to route the lightpath. The end-nodes of the lightpath access the lightpath with transmitters and receivers that are tuned to the wavelength on which the lightpath operates.

The wavelength-routed network has the following features:

- *Transparency:* Transparency refers to the property of a network for which the lightpaths can carry data at a variety of transmission speeds, formats, protocols, etc. and can be made protocol insensitive. In other words, it is the network property for which any signal travels along the network independent of its transmission format, speed, etc.
- *Scalability:* It is always possible to add more nodes to the network. Scalability dictates wavelength reuse. However, wavelength routing alone does not guarantee scalability. A high degree of scalability can be achieved when any channel can be switched to a new wavelength path, which may be at a different wavelength, thereby demanding wavelength translation.
- *Wavelength translation (or conversion):* Without wavelength translation, a lightpath must be assigned the same wavelength on all the links along its route. With wavelength translation, a lightpath can be assigned a different wavelength on different links along its route. Thus, wavelength translation can improve the utilization of the available wavelengths in the network.

4.5.2 Testbeds

Several wavelength-routed testbeds have been developed and built in the last few years.[38,39,44,45] These efforts have been funded by government agencies and interested telecommunication companies. MONET and RACE are two examples.

The Multiwavelength Optical NETworking (MONET) consortium was a 5-year program that started in December 1994. It was funded in part by the Defense Advanced Research Projects Agency (DARPA) and some participating companies — AT&T, Bell Atlantic (now Verizon), Telcordia Technologies (formerly Bellcore), Bell South, Pacific Telesis, and Lucent Technologies. Its objective was to prototype WDM optical networking technologies, network elements, and a network management system to support a testbed of wavelength-routed networks linking U.S. government agencies in the Washington, D.C. area. The testbed employed eight wavelengths spaced 200 GHz apart, with data transmission at 2.5 Gbps per wavelength over 2000 km of single-mode fiber. It integrated management of the optical, SONET, and ATM layers. MONET multiwavelength technology is being commercialized by the partner companies.

The RACE (Research and Development in Advanced Communication Technologies in Europe) program lasted from 1988 to 1995. It was sponsored by a consortium of industrial and academic groups in Europe and was implemented in two stages. RACE-I was established in 1988 to employ integrated broadband communication across Europe. The principal feature of this network structure is that the optical transport layer allows large blocks of capacity to be routed around the network without the need for optoelectronic conversion and processing. RACE-II was a follow-up program designed to move the results closer to implementation. It is a multi-wavelength transport network (MWTN) that explores the interaction between SDH (synchronous digital hierarchy) layer and the optical network layer. Both optical cross-connect (OXC) and optical add/drop multiplexer (OADM) were developed during the program. The testbed used four wavelengths spaced 4 nm apart, with each wavelength carrying SDH data at 622 Mbps or 2.5 Gbps. RACE-II provided a foundation for much of the work now being undertaken by ACTS (Advanced Communications Technologies and Services). The focus of ACTS is on implementing technology demonstrations in generic trials, while continuing to advance technology in areas that need further development. Its objective is to take RACE systems out of the laboratories and put them to test under real-world conditions in field trials across Europe.

4.6 Undersea networks

The underlying principles of optical fiber communications for inland applications are somewhat applicable to undersea and terrestrial systems. However, stringent reliability requirements must be imposed on the undersea networks because of the need to minimize the number of repairs during the life of the cable. The cable must have sufficient mechanical strength to survive without damage the harsh environment and the deep-sea cable-laying operations.

The explosive growth in demand for global telecommunication services has been witnessed in the 1980s and 1990s. Three primary forces are driving the expansion of the global telecommunication networks.[46] First, deregulation has increased the competition among service providers. Competition for telecommunication services has led to the building of the global infrastructure, including submarine cable systems. In 1966 alone, nearly 100,000 km of new cable were put into service worldwide. Deregulation has also increased opportunities for their services around the world. Consequently, data communications at the global level have become a fierce battleground for revenue generation. Second, computer networking has been recognized as vital to the success of companies and organizations. This recognition further fuels the keen competition among service providers who now perceive their revenues from voice service potentially at risk if they cannot satisfy an organization's overall voice and data communication needs. Third, many of the technical obstacles preventing the construction of high-speed integrated networks have been overcome. The basic technologies responsible for this change include digital switching, fiber optic transmission, and automated operation systems. These technologies are deployed worldwide and are changing the way WANs are built and operated. The change has greatly contributed to the rise of multinational corporations and the realization of a global economy.

It is hoped that the global network will provide reliable connectivity and quality services for new markets in emerging economies and newly opened markets previously served exclusively by state-run monopolies, and for meeting the increasing demands on existing traffic routes.[47] This situation has caused a rapid evolution in the undersea fiber-optic technology, which in just a couple of years has become the dominant means of intercontinental communication. The ultimate goal of the global telecommunication network is to usher in an era in which all coastal countries of the world will be connected in a vast, fast-growing, global undersea fiber-optic communication network.

4.6.1 Historical background

Undersea global communication has a rich history. The first transatlantic telegraph cable was successfully installed in 1858 after several attempted failures. The first analog Trans-Atlantic Transmission (TAT-1) coaxial cable went into service in 1956 with a capacity of 36 simultaneous voice-communications circuits or channels; it was taken out of service in 1979 when it had exceeded its 20-year design life.[48,49] Subsequently, several other analog systems were installed with capacities of 140, 840, and 4200 channels per cable. The last of these copper coaxial transatlantic cables and the last analog cable

Table 4.4 American Undersea Early (Coaxial) Cable History

1932	First telephone undersea cable between Key West, FL and Havana, Cuba
1952	Beginning of negotiations on TAT-1
1956	Scotland-Newfoundland (TAT-1) went into service
1957-1960	France-Newfoundland (TAT-2), Washington-Alaska, Florida-Puerto Rico, California-Hawaii (HAW-1)
1963	United States-United Kingdom (TAT-3)
1964	California-Hawaii (HAW-2), Hawaii, Japan (TP-1), Florida-St. Thomas (St. T-1)
1965	United States-France (TAT-4)
1968	Florida-St. Thomas (St. T-2)
1970	United States-Spain (TAT-5)
1974	California-Hawaii (HAW-3)
1975	Hawaii-Japan (TP-2)
1976	United States-France (TAT-6)
1982	Florida-St. Thomas (St. T-3)
1983	United States-United Kingdom (TAT-7)

(TAT-7) began service in 1983. A summary[48] of these early American undersea cables is given in Table 4.4. In only 20 years (1956–1976), the capacity of a single cable increased over 100-fold, from 36 to 4000 circuits. The revolutionary growth in cable capacities is due to progress in electronic technology.[50]

An historic milestone was reached in 1988 when the first undersea fiber-optic transatlantic telephone system (TAT-8) was installed. The total capacity of the TAT-8 system is 577 Mbps. This was also the longest, continuous digital transmission span ever installed. It linked the U.S., U.K., and France.[51] Soon after, the same technology was used to link the U.S. and Japan. This fiber-optic connection provided for the first time the prospect for seamless, high-capacity, digital networking among the world's three largest markets — the U.S., Europe, and Japan. The reaction from the business community was overwhelmingly positive with the result that TAT-8 reached its capacity within 18 months of installation.

The next transatlantic system (TAT-9) began service in early 1992 and spanned the U.S. and Canada to the U.K. The system's total available capacity is 16,000 channels, which can be expanded to 80,000 channels by digital-circuit multiplication techniques. The early TATs were designed and constructed as point-to-point systems between nodes of high-traffic demand. The newer systems (TAT-10 and TAT-11) were planned as networks with mutual restoration capability. The latest transatlantic systems are TAT-12 and TAT-13, which introduced the submerged erbium-doped fiber amplifiers (EDFAs) that enabled the installation of long unrepeatered systems. They are to provide services between the U.S., U.K., and France like the previous 11 TAT cables.[52,53] Transmitting at 5 Gbps, the two undersea systems provide a total capacity of over 300,000 circuits. Some of the recent American undersea cables are illustrated in Figure 4.21 and summarized in Table 4.5.

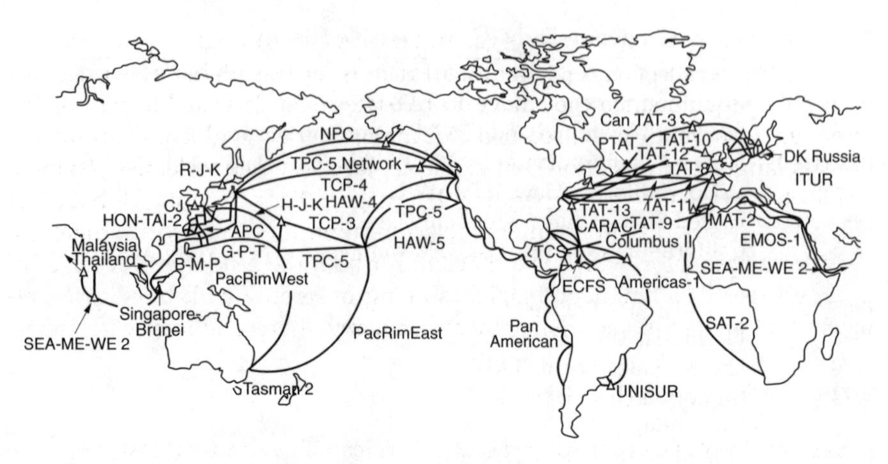

Figure 4.21 Global digital fiber optical network.

Table 4.5 American Undersea Recent (Fiber Optical) Cable History

1988	United States-United Kingdom and France (TAT-8) was installed
1992	United States-Canada and United Kingdom (TAT-9) began service
1993	TAT-10 and TAT-11
1992	Negotiations for TAT-12 and TAT-13 started
1995-1996	United States-United Kingdom and France (TAT-12/13) were installed

Although undersea coaxial cable and satellite communications made the world closer together, it was fiber-optic technology that brought today's global community within reach. And due to the worldwide concerted effort in photonics, breakthroughs have come one after another, resulting in a drastic decrease in the cost of cable systems.

4.6.2 Global network architecture

The global undersea communications networks can be viewed as consisting of three tiers: the domestic, regional, and interregional or global.[54,55] Most of these networks must be affordable, easy to install, highly reliable, easy to maintain and repair, and able to support traffic growth as long-term investments.

The *domestic network* connects population centers within a nation. It distributes traffic within that nation and aggregates outbound traffic to other nations.

The *regional network* connects nations within a geographic region and serves as a bridge between the domestic and global networks. It distributes traffic within the region and aggregates outbound traffic to other geographic regions. Most countries want to maintain sovereignty over their own traffic and do not want intrusion by neighboring nations. The sovereignty issue is resolved by placing the majority of the network on the sea bottom.

The global network connects geographic regions throughout the world — it spans the oceans that separate them. Cables for this global network must be highly reliable since they have to operate for a 25-year lifespan in the harsh environment on the sea bed.[56] A major barrier to global communications is language translation. For example, Japanese networks use Japanese Kanji characters for their operating language.[57] Because only alphabetical characters are used in international telecommunications networks, there is a need for a translator, which is inefficient for both the sender and receiver of the message. This language barrier is being overcome with intelligent networks that are readily accessible by anyone, anywhere, and in any language.

4.6.3 Africa ONE project

An example of a recent long-distance optical fiber undersea system, the Africa ONE project, is considered below.

It was recognized in 1993 that some parts of the world were not being included in the global undersea fiber-optic-based telecommunication networks. The continent of Africa was one of those regions. The International Telecommunication Union (ITU) approached AT&T to develop the Africa Optical Network (Africa ONE) project.[54,55]

The Africa ONE is a regional network that will encircle the entire continent with an undersea fiber-optic ring network. It is a telecommunications network capable of carrying voice, video, and data traffic. As shown in Figure 4.22, the network architecture is a hybrid self-healing ring that supports

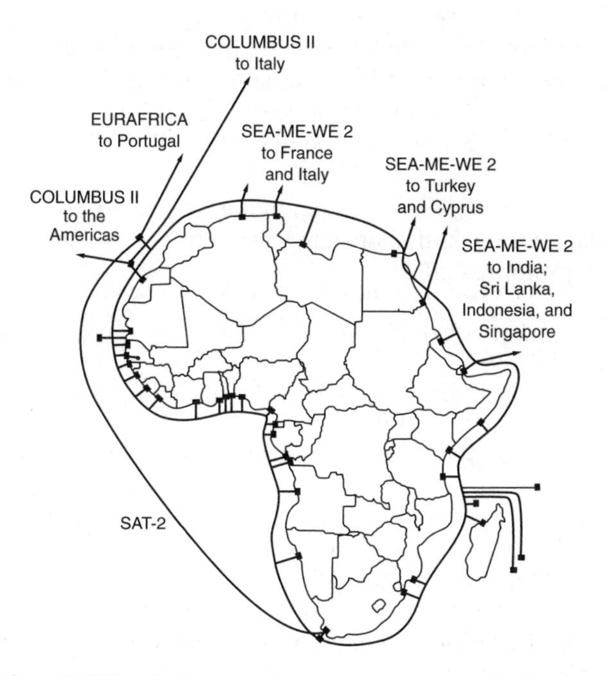

Figure 4.22 Africa ONE optical network.

a connectivity of 26 landing sites on the African coast as well as in southern Europe and the Middle East. The network will use 32,000 km of undersea fiber-optic cable and will facilitate the connection of all 54 African countries. It will be the largest network ever conceived. Inland countries lacking coastal access can have their traffic merged with the network through an interregional gateway of their choice. The network will do more for Africa than meeting telecommunication needs. It is hoped that the network will bring advances in education and stimulate economic development that will generate jobs.

Africa has limited direct telecommunication connectivity to the rest of the world. African teledensity is very low (roughly six lines per 1000 people), and the technology supporting telecommunications is poorly maintained and outdated. Each African nation has its own telecommunication authority, and the continent is presently united by satellite and microwave communications administered by international authorities: the Regional African Satellite Communications Organization (RASCOM) and the Pan African Telecommunications Union (PATU). In spite of the two international authorities, about 90% of intra-African telephone traffic is routed via non-African administrations. Most international traffic is routed through Europe. Consequently, huge transmit fees are paid to foreign correspondents. Africa ONE was developed with those problems in mind.

Special requirements for Africa ONE include the following:[54]

- *Connectivity:* It must connect all African coastal nations and interconnect the landlocked nations to the network. The hybrid ring nature of the network meets this requirement.
- *Sovereignty and security:* It must be capable of allowing traffic directed to a specific nation to be terminated only in that intended nation. Wavelength-division multiplexing (WDM) is used to support this.
- *Fault tolerance:* It must be able to protect all traffic paths in the event of any single fault.
- *System flexibility:* Adding/removing traffic to/from the network should be easy, and connecting new nations to the network should require minimal effort.

In addition, the network must meet the telecommunication needs for the next 25 years. It is designed to be managed, operated, and maintained in Africa. Finally, it must be compatible with International Telecommunication Union (ITU) standards.

The Africa ONE project will be jointly constructed by AT&T Submarine Systems, Inc. and Alcatel Submarine Networks at an estimated cost of $2.6 billion. It will be ready for service in 2002. Once Africa ONE is deployed, it will have far-reaching benefits for African nations.[58]

The level of undersea networking will definitely increase. The actual direction of undersea systems will be dictated by technological breakthroughs in the coming years. It is expected that the system will be transmitting at rates 20 Gbps or higher.

4.7 Emerging technologies

This last section gives a brief overview of other protocols and technologies that will likely contribute to the solution of capacity, multicast, scalability, quality of service, and other issues affecting next generation networks. These technologies are still evolving.

4.7.1 Optical gigabit Ethernet

While the deployment of fiber optics is extending from the backbone to the WAN and MAN and will soon penetrate into the local loop, Ethernet is spreading from the LAN arena to the MAN and WAN as the uncontested standard.

Ethernet was defined by the IEEE 803 standard as the shared medium for LANs using a distributed medium access control mechanism known as carrier sense multiple access with collision detection (CSMA/CD). Ethernet has become the *de facto* LAN standard. More than 90% of today's LANs are Ethernet based, and the market share is increasing. Quarter-century-old Ethernet remains one of the most cost-effective, scalable, and handy networking technologies available. Optical Ethernet advances Ethernet features to levels undreamed of and not even feasible with copper-based technologies.

Optical Ethernet is the technology that extends Ethernet beyond the local area networks (LANs) and into metropolitan area networks (MANs) and wide area networks (WANs). While Ethernet LANs are used within the enterprise, optical Ethernet technology can be used as a service-provider offering. Optical Ethernet does not have physical distance limitations as does the native Ethernet; it can span sites miles apart. It comes with a set of desirable network services such as dynamic host configuration protocol (DHCP), domain name system (DNS), and traffic encryption. Transport is straightforward and essentially a layer-2 function. This transport style has the advantage of simplicity and ease of integration with long-haul DWDM systems. MANs based on optical Ethernet offer a high-bandwidth, low-cost, easy-to-learn-and-manage alternative to conventional MAN technologies such as SONET.

There are three possible ways of extending Gigabit Ethernet into the MAN/WAN. Using Ethernet

- on dark fiber
- over all-optical DWDM networks
- over SONET

Each of these approaches has its merits and disadvantages. Gigabit Ethernet on dark fiber is perhaps the least expensive and most practical solution. However, it is practical only for distances up to 1000 km and up to eight wavelengths. All-optical networks do not require any electronic repeater equipment and can carry any data format whether it is Gigabit Ethernet,

ATM, or SONET. However, all-optical networks are optimized for the SONET signaling rates of 2.5 and 9.98 Gbps. To insert 1.25 or 12.5 Gbps Ethernet signaling may be inefficient. Ethernet over SONET involves encapsulating Ethernet frames into SONET frames. This approach is attractive to carriers because they can continue to leverage the investment in SONET networks. As mentioned earlier, SONET networks will be inefficient for carrying Ethernet, particularly 10 Gigabit Ethernet, because its signaling does not match the standard SONET OC-n signaling rates.

The drive for optical Ethernet is motivated by at least three powerful forces. First, there is the need for integration of services, and there are a number of limitations to the ATM/SONET architecture. A few years ago, it was thought WAN technologies such as ATM would be the unifying protocol to integrate LAN and WAN services — from the desktop to the backbone. However, ATM has largely failed to penetrate the LAN market because of its complex management and cost. In the same way, SONET is inefficient for data transmission because it offers time slots for individual streams of incoming traffic rather than a big pipe, and it lacks a native interface for IP. Ethernet-based Internet access is a more cost-effective topology. For example, an OC-48 SONET port running at 2.4 gigabit costs roughly $30,000 while a 1-gigabit Ethernet port costs around $1200. Only large enterprises can afford to spend $3200 to $4300 per month to lease a T-3 (45 Mbps) or OC-3 (155 Mbps). Second, the huge economies of large scale and mass market of the Ethernet technology are rapidly bringing down the cost of Ethernet switches and hubs. The low cost and simplicity of Ethernet technology has resulted in an explosion of LAN technologies and made high-speed variations of Ethernet ready to invade the WAN. Third, optical-Ethernet networking combines the ubiquity of IP protocol, the low cost of fiber optics, and the reliability of high-speed Ethernet. Competitively important for next generation carriers, optical Ethernet also reduces the amount of both equipment and power needed at the central office.

The concept of optical Ethernet is evolving from the same place as the Internet evolved — the university and research community. However, due to regulatory and competitive restraints in the WAN market, it will take time for optical Ethernet to penetrate the market.

4.7.2 DTM

Computer networks, such as the Internet, have traditionally provided asynchronous communication, which uses packet-switching and store-and-forward techniques. Telephone networks, on the other hand, have provided real-time communication using circuit-switching and time-division techniques. Asynchronous transfer mode (ATM) has been introduced to provide all kinds of services. However, it has been clear that ATM cannot cope with services that require high quality of service (QoS) guarantees such as voice, video, and hi-fi audio in an economically satisfactory manner. IP over ATM has been found to be a less elegant combination of protocols. There is a need to find a transfer mode that can accommodate real-time as well as asynchronous traffic. Also, with the large amount of data transfer capacity offered by the current fiber networks, processing and buffering at switch and access points on the network are causing a bottleneck problem. Against this background, dynamic synchronous transfer mode (DTM) was developed.[59–61]

DTM is a new transport network technology designed specifically for the foreseen explosion of real-time media in the next generation networks. It is a broadband network architecture developed at the Royal Institute of Technology, Sweden. It is an attempt to combine the advantages of both asynchronous and synchronous media access schemes. It is a networking scheme designed to utilize fully the capacity of optical fiber as a physical medium by emphasizing simplicity and avoiding computation-intensive policing, queuing, buffering, and control mechanisms. This goal is achieved through the technology's inherent characteristics, which include dynamic bandwidth allocation, low propagation delay, almost zero delay variation, full traffic isolation between channels, and high-speed transmission. A standardization effort on DTM is still in progress.

DTM was developed to meet layer-2 requirements for advanced IP services. Although layer-2 requirements can be rather extensive, they include:[62]

- *Simple scalability:* As the demand on networks has increased, the need for easily scalable resources has grown exponentially.
- *High availability:* To ensure high availability, the network must be designed with high link, node, and switching redundancy at low cost.
- *Support for integrated services:* To ensure a high-quality IP network, layer-2 platform should provide real-time support with nearly jitter-free transport.
- *Resource management:* The resource management feature provides the network operator with billing information.

DTM is based on circuit switching augmented with dynamic reallocation of time slots. It is basically a time division multiplexing (TDM) scheme. It is a switching as well as a transmission scheme that can serve as a substitute for ATM. DTM is similar to SONET/SDH, and it can run on top of SONET/SDH or work stand-alone. It is compared with other transport network technologies in Figure 4.23. DTM guarantees high transport quality even over large-scale networks, and it works well over DWDM.

DTM uses a unidirectional medium with capacity shared by all connected nodes, i.e., multiple access. The medium (called *link*) can assume different topologies, such as dual bus, folded bus, or ring. Since DTM is based on time division multiplexing (TDM), the entire capacity of a fiber link is divided into small fixed-size 125-μs frames. Each frame is further divided into 64-bit time slots. The frame length of 125 μs and 64 bits per slot is to enable easy adaptation to digital voice. The number of slots per frame depends on the bit rate. For example, a bit rate of 2.5 Gbps produces approximately 4800 slots per frame. As shown in Figure 4.24, a slot is either a control slot

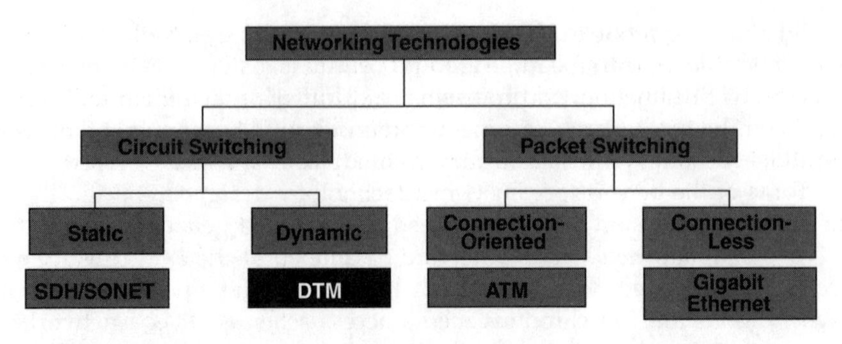

Figure 4.23 Comparison of network technologies.

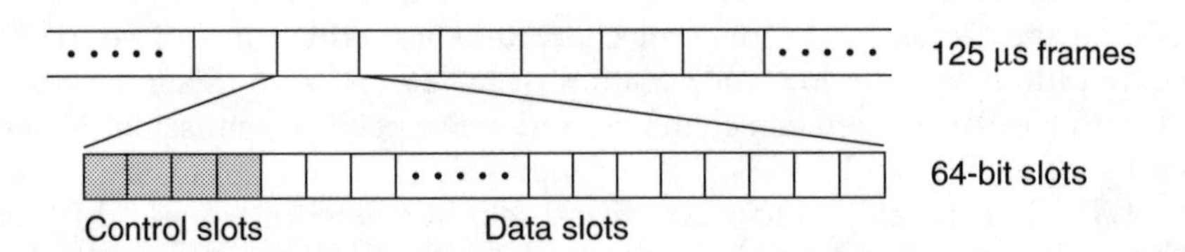

Figure 4.24 DTM multiplexing format.

or data slot, and, if necessary, control slots may be converted to data slots or vice versa. The majority of the slots are data slots with payload. At initialization, data slots are allocated to nodes according to a predefined distribution; a node needs to own a slot to transmit. DTM employs a distributed algorithm for slot reallocation, where free slots are distributed among the nodes.

The service provided by a DTM network is based on channels. A channel consists of a set of time slots with a sender and an arbitrary number of receivers. It gives guaranteed capacity and constant delay, making it suitable for delay-sensitive applications with real-time requirements, such as video and audio. Channels on the physically shared medium are realized by a time division multiplexing (TDM) scheme, as shown in Figure 4.24. A node forms a channel by allocating a set of data slots for the channel and by sending a control message for channel establishment. A channel can take one of the following forms:

- *Simplex:* set up from sender to receivers
- *Duplex:* consists of two channels, one in each direction
- *Multirate:* consists of an arbitrary number of data slots
- *Multicast:* has several receivers

Traditionally, a circuit is a point-to-point connection between a transmitter and a receiver. DTM employs a shared medium that inherently supports multicast since a slot can be read by several nodes on a bus.

DTM is a network architecture for integrated services. It offers different types of traffic: IP over DTM, DTM LAN emulation (DLE), PDH transport, SDH/SONET tunneling, and transport of studio video. DLE is used to connect LANs by transferring Ethernet packets over DTM channels and making the attached nodes appear as if they were a part of a LAN.

Some of the benefits for DTM include:

- High bandwidth utilization and fragmentation of total capacity
- Customized billing
- Scalable in space and capacity
- Secure transport of confidential information
- Support of integrated services
- High availability

DTM avoids the major drawbacks of ATM. First, space overhead is low for DTM because it uses time-division switching, so no packet header is needed. In contrast, space overhead for ATM is about 10% ($100 \times 5/53$). Second, DTM provides good QoS because it is based on TDM. Also, because DTM is based on circuit-switching, traffic management is simplified. In contrast, ATM is based on statistical multiplexing, so cell loss and cell transfer delay jitter may occur. Third, a DTM network uses source-routing and on-the-fly connection setup, i.e., connections are created and released dynamically on a burst-by-burst basis. This aspect makes DTM suitable for transferring fairly large data, such as the World Wide Web (WWW) traffic, because such transfer does not require a pre-established connection. ATM is not suited to WWW traffic because it is connection oriented and it incurs a large connection setup delay when the network becomes large.

4.7.3 MPLS

With the rapid growth of Internet services and the recent advances in dense wavelength division multiplexing (DWDM) technology, an alternative to ATM for multiplexing multiple services over individual circuits is needed. The once fast and high bandwidth ATM switches are not good enough as they are being out-performed by Internet backbone routers. Multiprotocol label switching (MPLS) offers a simpler mechanism for packet-oriented traffic engineering.

MPLS is an extension to the existing IP architecture and the latest step in the evolution of routing and forwarding technology for the Internet core. It is a new technology being standardized by the Internet Engineering Task Force (IETF) designed to enhance the speed, scalability, and service provisioning capabilities of the Internet. As a technology for backbone networks, MPLS can be used for IP as well as other network-layer protocols. It has also become the prime candidate for IP-over-ATM backbone networks.

The aim of MPLS is to improve the scalability and performance of the prevalent hop-by-hop routine and forwarding across packet networks. Its primary goal is to standardize a technology that integrates the label switching forwarding paradigm with network layer routing.

MPLS is based on the following key concepts:[63–65]

- It provides a means to map IP addresses to simple, fixed-length, protocol-specific identifiers known as "labels," i.e., it separates forwarding information (label) from the content of the IP header. The labels are used by different packet-forwarding and packet-switching technologies.
- It uses a single forwarding paradigm (label swapping) at the data plane to support multiple routing paradigms at the control plane.
- It is independent of the underlying link-layer technology and uses different technologies and link-layer mechanisms to realize the label swapping forwarding paradigm. It supports IP, ATM, and frame-relay layer-2 protocols.
- It provides flexibility in the delivery of new routing services.
- It supports the delivery of services with QoS guarantees.

Label switching is key to MPLS. At the ingress point of an MPLS network, the *label edge router* (LER) adds a label to each incoming IP packet according to its destination and the state of the network. Label switching relies on the setup of switched paths through the network, which are called *label switching paths* (LSPs). In other words, LSP is an established logical MPLS connection that links a LER via a *label switched router* (LSR) to another LER, as shown in Figure 4.25. When a packet is sent on an LSP, a label is applied to the packet, as shown in Figure 4.26. On ATM links, for example,

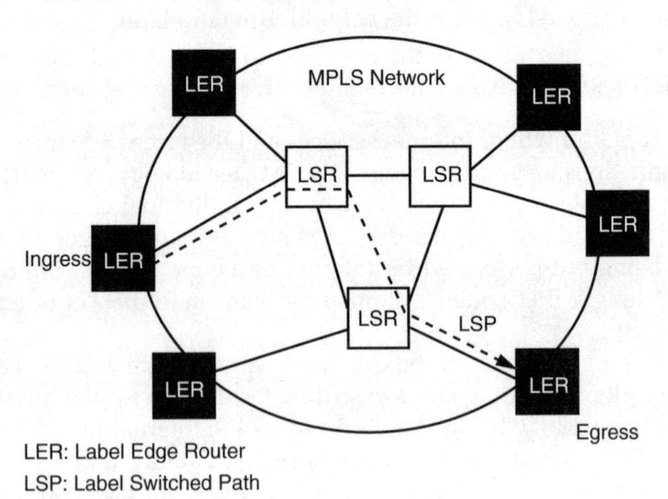

Figure 4.25 MPLS architecture.

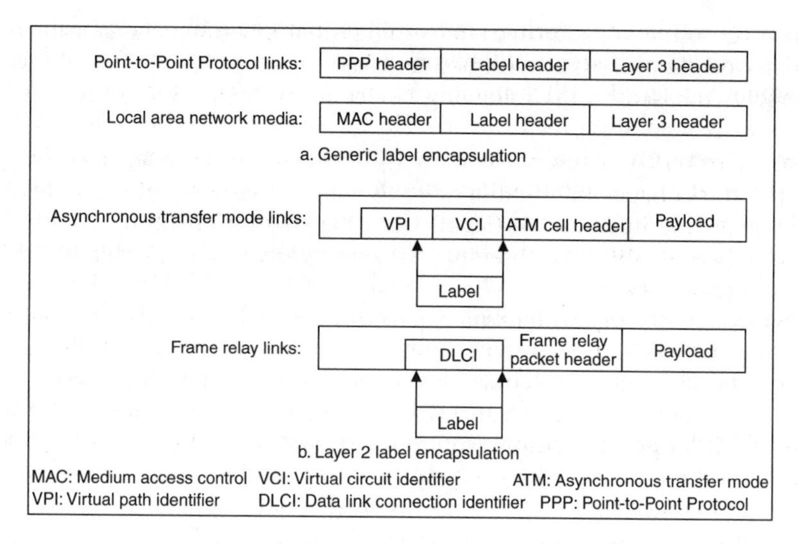

Figure 4.26 Label formats for different packets.

the label may be carried as the virtual circuit identifier and/or virtual path identifier applied to each ATM cell. A similar scheme has been proposed for SONET.

In an MPLS network, an LSP is created for each path through the network. The router examines the header to determine which LSP to use, adds the appropriate LSP identifier (a label) to the packet, and forwards it to the next hop. All the subsequent nodes forward the packet along the LSP identified by the label. LSPs typically follow the shortest path from source to destination. Once the traffic is within the MPLS network, only the label is used to decide to which hop the packet is to be sent. Each label has only hop-by-hop relevance. This feature allows for efficient resource usage and provides high speed forwarding.

The main component of MPLS architecture is the label distribution protocol (LDP), which is a set of protocols by which a label switch router (LSR) communicates with an adjacent peer by means of exchanging labels. Each LSP has a forwarding equivalent class (FEC) associated with it. The classification of packets into specific FECs identifies the set of packets that will be mapped to a path through the network. At the ingress of an MPLS domain, all incoming packets are assigned to one of these FECs. The LDP protocol has four types of messages: discovery, session, advertisement, and notification.[66]

MPLS provides a variety of benefits to service providers. First, it supports virtual private network (VPN) services. The use of MPLS for VPNs is an alternative to building VPNs with ATM or frame-relay permanent virtual circuits (PVCs). Second, MPLS is useful in field-of-traffic engineering, which refers to the ability to control traffic flows in a network with the goal of reducing congestion and improving network utilization. In data networks, the MPLS traffic engineering control block performs all crucial functions, such as resource discovery, network state, path computation, and route management. Third, MPLS enables IP and ATM integration. In fact, MPLS is primarily a solution for IP-over-ATM backbones. MPLS enables an ATM switch to perform virtually all of the functions of an IP router.

The next logical evolutionary step toward all-optical networks, already under investigation, is to employ a variation of MPLS that supports different wavelengths for different data flows on the transport layer. This technique is called generalized multiprotocol label switching (GMPLS), which is also known as multiprotocol lambda switching (MPλS).[67] GMPLS is a suite of protocol extensions that provides common control to packet, TDM, wavelength, and fiber services. It extends the MPLS and LSP mechanisms to create generalized labels and generalized LSPs. By consolidating different traffic types, GMPLS permits simplification of networks and improves their scalability. It offers the means by which networks can be scaled and simplified by deployment of a new class of network element designed to handle multiple traffic types simultaneously. A key benefit of GMPLS is that it gives network operators the freedom to design their networks to best meet their specific objectives.

The work on MPLS continues. The MPLS Forum, an international organization (www.mplsforum.org), advances the successful deployment of multivendor MPLS networks. It is hoped that optical MPLS over DWDM will also emerge from the MPLS standards.

Summary

The increasing demand for bandwidth, changes in the regulatory environment, and introduction of new services will have a profound impact on the architecture and functionality of next generation optical networks.The ultimate goal of next generation networks is for information to be able to reside anywhere and to be accessible from everywhere as if it were located locally. Achieving that goal requires high-capacity optical networks and their rapid transition from research laboratories to commercial deployment.

This chapter has explored optical networks for local, metropolitan, and global coverage. Such networks include FDDI, SONET, fiber channel, broadcast-and-select networks, wavelength-routed networks, and undersea networks. Optical WDM networks are capable of fulfilling the enormous bandwidth demands of present and future applications. This goal is achieved with new optical components for routers, cross-connects, and add/drop multiplexers. One can expect the capacity per fiber to increase from the present Tbps level to some 100 Tbps before fundamental limits inhibit progress, if at all.

The chapter has also considered emerging technologies such as optical Ethernet, dynamic synchronous transfer mode (DTM), and multiprotocol label switching (MPLS). The optical Ethernet architecture promises to be the dominant means of integrating data, voice, and video services over a single network in a cost-effective manner. DTM is designed to utilize fully the almost unlimited capacity of optical fibers. MPLS came of the necessity to address new connection-oriented needs of the Internet. It provides the ability to support any type of traffic on a large IP network without subjecting the design to the limitations of different routing protocols, transport layers, and addressing schemes. With these emerging technologies, the stage is set for a paradigm shift in the communications industry that could well result in completely new equipment deployment grounded in the wide adoption of fiber optics and Ethernet technologies.

References

1. M. N. O. Sadiku, *Wide and Metropolitan Area Networks,* Prentice-Hall, Upper Saddle River, NJ, in press.
2. S. P. Joshi, High-performance networks: A focus on the fiber distributed data interface (FDDI) standard, *IEEE Micro,* vol. 6, no. 3, June 1986, 8–14.
3. R. Jain, *FDDI Handbook: High-Speed Networking Using Fiber and Other Media,* Addison-Wesley, Reading, MA, 1994.
4. A. Shah and G. Ramakrishnan, *FDDI: A High Speed Network,* Prentice-Hall, Englewood Cliffs, NJ, 1993.
5. W. E. Burr, The FDDI Data Optical Link, *IEEE Commun. Mag.,* vol. 24, no. 5, May 1986, 8–23.
6. F. E. Ross, FDDI — a tutorial, *IEEE Commun. Mag.,* vol. 24, no. 5, May 1986, 10–17.
7. F. E. Ross, An overview of FDDI: The fiber distributed data interface, *IEEE J. on Selected Areas in Comm.,* vol. SAC-7, no. 7, Sept. 1989, 1043–1051.
8. D. Tsao, FDDI: Chapter two, *Data Commun.,* Dec. 21, 1991, 59–70.
9. G. Watson and D. Cunningham, FDDI and beyond: A network for the 90s, *IEE Rev.,* April 1990, 131–134.
10. J. F. Mazzaferro, An overview of FDDI, *J. Data Comput. Commun.,* vol. 3, no. 1, 1990, 15–27.
11. K. Restivo, The boring facts about FDDI, *Data Commun.,* vol. 23, no. 18, 1994, 85–90.
12. P. Davids et al., FDDI: status and perspectives, *Comput. Networks ISDN,* vol. 26, no. 8, 1994, 657–677.
13. American National Standard, Fiber distributed data interface (FDDI) token ring media access control (MAC), ANSI Standard X3.139, 1987.
14. American National Standard, Fiber distributed data interface (FDDI) physical layer protocol (PHY), ANSI Standard X3.148, 1988.
15. American National Standard, Fiber distributed data interface (FDDI) physical layer medium dependent (PMD), ANSI Standard X3.166, 1990.
16. American National Standard, Fiber distributed data interface (FDDI) station management (SMT), ANSI Standard X3.229, 1994.
17. American National Standard, Fiber distributed data interface (FDDI) hybrid ring control (HRC), ANSI Standard X3.186, 1992.
18. G. C. Kessler and D. A. Train, *Metropolitan Area Networks,* McGraw-Hill, New York, 1992.
19. G. C. Kessler, Simplifying SONET, *LAN Mag.,* July 1991, 36–46.

20. T. Russell, *Telecommunications Protocols,* McGraw-Hill, New York, 1997, 337–355.
21. H. Fujita, K. Sakai, and T. Aoyama, SONET fiber-optic transmission systems, *Hitachi Rev.,* vol. 44, no. 4, 1995, 187–192.
22. J. Cashin, *High-Speed Networking,* Computer Technology Research Corp., Charleston, SC, 1995, 67–75.
23. G. C. Kessler, An overview of the synchronous optical network, in G. R. McClain (Ed.), *Handbook of Networking and Connectivity,* AP Professional, Boston, 1994, 139–165.
24. T. Beninger, *SONET Basics,* Telephony, Chicago, IL, 1991, 17–23.
25. S. Kapoor, Technology update: Synchronous optical networks (SONET), *New Telecom Q.,* vol. 2, 1994, 53–56.
26. R. J. Krajcik and K. E. Skoglund, SONET: New standard for high-speed data networks, *AT&T Technol.,* vol. 8, no. 1, 1993, 18–20.
27. C. Clendening, C. Harris, and A. Farinholt, SONET network evolution toward ATM in the USA, *Fujitsu Sci. Tech. J.,* vol. 32, no. 1, June 1996, 13–35.
28. W. J. Goralski, *SONET: A Guide to Synchronous Optical Network,* McGraw-Hill, New York, 1997, 388–408.
29. I. Haque, W. Kremer, and K. Raychaudhuri, Self-healing rings in a synchronous environment, *IEEE LTS,* November 1991, 30–37.
30. *IEEE Commun. Mag.,* August 1990, is a special issue on SONET/SDH.
31. *IEEE LTS Mag.,* vol. 3, no. 4, November 1991, is a special issue on SONET.
32. A. F. Benner, *Fibre Channel,* McGraw-Hill, New York, 1996, 6.
33. A. X. Widmer and P. A. Franaszek, A DC-balanced, partitioned block, 8B/10B transmission code, *IBM J. Res. Develop.,* vol. 27, no. 5, Sept. 1983, 440–451.
34. P. A. Franaszek and A. X. Widmer, Byte oriented DC balance (0,4) 8B/10B partitioned block transmission code, U.S. Patent No. 4,486,739, Dec. 4, 1984.
35. D. Getchell and P. Rupert, Fiber channel in the local area network, *IEEE LTS,* May 1992, 38–42.
36. G. R. Stephens and J. V. Dedek, *Fibre Channel: The Basics,* ANCOT Corporation, Menlo Park, CA, 1997, 3–5.
37. A. Anzaloni et al., Fiber channel FCS/ATM interworking: design and performance study, *Proc. IEEE GLOBECOM,* vol. 3, 1994, 1801–1807.
38. R. Ramaswami and K. N. Sivarajan, *Optical Networks: A Practical Approach.* Morgan Kaufmann, San Francisco, 1998, 291–328, 463–480.
39. B. Mukherjee, *Optical Communication Networks,* McGraw-Hill, New York, 1997, 109–218, 367–394.
40. L. Kazovsky et al., From STARNET to CORD: Lessons learned from Stanford WDM projects, in G. Prati (Ed.), *Photonic Networks,* Springer-Verlag, London, 1997, 300–330.
41. E. Hall et al., The Rainbow-II gigabit optical network, *IEEE J. Selected Areas Commun.,* vol. 14, no. 5, June 1996, 814–823.
42. F. J. Janniello, R. Ramaswami, and D. G. Steinberg, A prototype circuit-switched multi-wavelength optical metropolitan-area network, *J. Lightwave Tech.,* vol. 11, no. 5/6, May/June, 1993, 777–782.
43. T. K. Chiang et al., Implementation of STARNET: A WDM computer communications network, *IEEE J. Selected Areas Commun.,* vol. 14, no. 5, June 1996, 824–839.

44. S. R. Johnson and V. L. Nichols, Advanced optical networking — Lucent's MONET network elements, *Bell Labs Tech. J.*, Jan.-Mar. 1999, 145–162.
45. B. Fabianek, Optical network research and development in European Community programs: from RACE to ACTs, *IEEE Commun. Mag.*, April 1997, 50–56.
46. C. Hemrick, Building today's global computer internetworks, *IEEE Commun. Mag.*, Oct. 1992, 44–49.
47. T. Soja, Crosscurrents and opportunities: the undersea fibre-optics industry, *Telecommunications* (International edition), vol. 30, no. 1, Jan. 1996, 5 pp.
48. R. D. Ehrbar, Undersea cables for telephony, in P. K. Runge and P. R. Trischitta (Eds.), *Undersea Lightwave Communications*, IEEE Press, New York, 1986, 3–22.
49. *Bell System Tech. J.*, vol. 36, no. 1, Jan. 1957.
50. J. H. Davis, N. F. Dinn, and W. E. Falconer, Technologies for global communications, *IEEE Commun. Mag.*, Oct. 1992, 35–43.
51. P. K. Runge, Undersea lightwave systems, *AT&T Tech. J.*, vol. 71. Jan./Feb. 1992, 5–13.
52. P. Trischitta et al., The TAT-12/13 cable network, *IEEE Commun. Mag.*, Feb. 1996, 24–28.
53. C. Reinaudo, Undersea cables: A state-of-the-art technology, *Electr. Commun.*, 1994, 5–10.
54. W. C. Marra and J. Schesser, Africa ONE: The Africa Optical Network, *IEEE Commun. Mag.*, Feb. 1996, 50–57.
55. J. C. Zsakany et al., The Application of undersea cable systems in global networking, *AT&T Tech. J.*, vol. 74, Jan./Feb. 1995, 8–15.
56. C. Reinaudo, Cable for submarine telecommunication systems, *Elect. Commun.*, vol. 63, no. 3, 1989, 226–230.
57. K. Tomaru et al., Global corporate networks in Japan, *IEEE Commun. Mag.*, Oct. 1992, 64–69.
58. For an update on Africa ONE project, check www.africaone.com/english.
59. C. Bohm et al., The DTM gigabit network, *J. High Speed Networks*, vol. 3, no. 2, 1994, 109–126.
60. N. Yamanaka and K. Shiomoto, DTM: Dynamic transfer mode based on dynamically assigned short-hold time-slot relay, *IEICE Transactions on Communication*, vol. E82-B, no. 2, Feb. 1999, 439–446.
61. C. Bohm et al., Fast circuit switching for the next generation of high performance networks, *IEEE J. Selected Areas Comm.*, vol. 14, no. 2, Feb. 1996, 298–305.
62. O. Schagerlund, Understanding dynamic synchronous transfer mode technology, *Comput. Technol. Rev.*, vol. 18, no. 3, March 1999, pp. 32, 36, 54.
63. J. Lawrence, Designing multiprotocol label switching networks, *IEEE Commun. Mag.*, July 2001, 134–142.
64. A. Viswanathan et al., Evolution of multiprotocol label switching, *IEEE Commun. Mag.*, May 1998, 165–173.
65. G. Hagard and M. Wolf, Multiprotocol label switching in ATM networks, *Ericsson Rev.*, vol. 75, no. 1, 1998, 32–39.
66. F. Holness and J. Griffiths, Multiprotocol label switching within the core network, *Proc. 38th European Telecommun. Cong.*, 1999, 97–100.
67. A. Banerjee et al., Generalized multiprotocol label switching: an overview of signaling enhancements and recovery techniques, *IEEE Commun. Mag.*, July 2001, 144–151.

Problems

4.1 Discuss the four core standards that make up FDDI.

4.2 It has been found that it could take time

$$t = (N-1)T + 2\tau$$

before a station can access the FDDI ring, where N is the number of stations, T = TTRT, and τ = ring latency. If a ring has 51 stations with T = 100 ms, and τ = 1 ms, how long does it take before each station could transmit? Is this time acceptable for isochronous traffic?

4.3 (a) Show that for large N, Equation 4.4 becomes

$$\eta_{max} = 1 - \frac{\tau}{T}$$

(b) An FDDI network has T = 8 ms, and τ = 1 ms. Find τ for 20 stations and for 100 stations.

4.4 Calculate the ring latency for these cases:
(a) a typical network with 30 stations and a total of 5 km of fiber in the ring
(b) a big network with 150 stations attached to a ring with 200 km of fiber
(c) a large network with 1000 stations attached to 200 km of fiber
Take the token processing time as 1 μs in all cases.

4.5 (a) Discuss the concept of "target token rotation time" in FDDI.
(b) How is transmitting synchronous traffic different from transmitting asynchronous traffic in FDDI?

4.6 Differentiate between synchronous, asynchronous, and isochronous traffic. Explain how each is supported by FDDI-II.

4.7 What are the differences between FDDI and FDDI-II?

4.8 An FDDI ring is 20 km long with end-to-end propagation delay of 118 μs latency of 0.4 μs per station. Calculate the number of stations connected to the ring. Take u = 2×10^6 m/s.

4.9 The name "fiber distributed data interface" (FDDI) has become a misnomer. Why?

4.10 Discuss the STS-1 frame in SONET.

4.11 Show that the bit rate of the STS-1 frame is 51.84 Mbps.

4.12 What are the protocol layers in SONET?

4.13 Bandwidth to be transported across the SONET ring is assigned to each customer in T1, T3, or OC-*n* sized blocks. An OC-1 carries an equivalent of how many T1 links? A 10-Mbps LAN requires a T3-sized chunk of bandwidth to transverse a SONET ring. How much bandwidth is wasted?

4.14 Explain the SONET concepts of add/drop multiplexer, terminal multiplexer, and digital cross-connect system.

4.15 What are the limitations of ATM/SONET architecture?

4.16 Explain why the term "fiber channel" is a misnomer.

4.17 Discuss the advantages and disadvantages of broadcast-and-select networks.

4.18 What is wavelength-discontinuity constraint?

4.19 Discuss the barriers of global undersea networks.

4.20 Compare and contrast ATM and DTM.

4.21 What is the Africa ONE project all about?

4.22 What is optical Ethernet? Discuss different ways it can be implemented.

4.23 What are the factors motivating the development of optical Ethernet?

4.24 How does DTM avoid the major drawbacks of ATM?

4.25 What is MPLS? What problem was it designed to solve?

part 2

Wireless networks

chapter five

Fundamentals of wireless networks

Any sufficiently advanced technology is indistinguishable from magic.

— Arthur C. Clarke

We live in the Information Age, an age of constant change, driven by our growing dependence on computers, cellular phones, pagers, fax machines, e-mail, and the Internet. This demand for real-time information exchange is made by an increasingly mobile workforce. Many jobs require workers to be mobile, e.g., inventory clerks, healthcare workers, policemen, and emergency care specialists. It is estimated that roughly 48 million U.S. workers cannot do without a form of wireless communication. Their work tools must be mobile.

Wireless communications is one of the fastest growing fields in engineering. Wireless technologies are targeted to certain needs, including:[1]

- Bypass of physical barriers (roads, railroads, buildings, rivers, etc.)
- Remote data entry
- Mobile applications (car, airplanes, etc.)
- Connectivity to hard-to-wire places
- World-wide connectivity for voice, video, and data communications
- Satellite communications
- Inexpensive setup

As in wired networks, there are three kinds of wireless networks:

- Wireless local area networks (WLAN), which provide users with mobile access to a wired LAN in its coverage area
- Wireless metropolitan area networks (WMAN), which are packet radio systems often used for law enforcement or utility applications
- Wireless wide area network (WWAN), which transmit data over cellular or packet radio

These networks are characterized the same way as their wire-bound counterparts. The IEEE 802 Standards Committee is developing standards for these networks. These wireless networks evolve around new and old technologies, including:[2]

- Mobile radio
- Spread-spectrum radio
- Cellular radio
- Infrared communications
- Meteor burst communications
- Mobile satellite communications
- Microwave communications
- Very small aperture terminal (VSAT)
- Personal communications systems (PCSs) and personal communications networks (PCNs)

Wireless communication works on the same set of fundamental principles, whether it is computer nodes on a LAN or the common cordless telephone. This chapter examines those common fundamentals, discussing the basic concepts of analog and digital mobile, wireless, and personal communications systems (PCSs). We begin with a brief history of wireless communication, which should help the reader appreciate the tremendous impact wireless communication will have on the way people live for the next several decades. We then discuss some fundamental concepts in wireless communication systems such as wave propagation, modulation schemes, and multiple access techniques.

5.1 A glimpse of history

Wireless communications is not new; it has been around for decades. Heinrich Hertz, Nikola Tesla, Guglielmo Marconi, and others experimented with the transmission and reception of radio waves in the late 19th century. The actual birth of radio occurred in 1897 when Marconi first demonstrated that radio could provide wireless communication between ships. Since then, the ability to communicate with "nomadic" people has increased tremendously.

The development of wireless communications can be regarded as taking place in three phases. Between 1907 and 1945 is the pioneer phase. In 1907, Lee De Forest invented the triode, which made possible for the first time both amplitude modulation (AM) and amplification of weak radio signals. A war-time ban on nonmilitary broadcasting delayed the acceptance of radio until the ban was lifted in 1919. Thereafter, hundreds of amateur stations sprang up. Edwin Armstrong invented frequency modulation (FM) in 1935. World War II was a stimulus to wireless communications; it led to the development of consumer radio and television systems after the war.

Between 1946 and 1968 is the initial commercial phase. Although the first regular commercial radio broadcast began in 1920, the golden age of broadcasting is generally considered to be from 1925 to 1950. The development of the transistor helped to reduce dramatically the size and power consumption of radio systems. Television then became the major electronic entertainment in the 1950s. The American people became fascinated with having TV at home; by 1950, there were over 104 TV stations broadcasting to over 6 million sets. About 750 million people watched the funeral of President John F. Kennedy in 1963, and about 600 million people saw man first set foot on the moon in 1969. With the advent of TV, the demise of radio was predicted by many experts. However, the medium flourished, and the number of stations multiplied, growing to over 4900 FM and 4200 AM stations by 1990.

The third and perhaps the most exciting phase began in 1969 and continues today. This phase includes the beginning of cellular, mobile, satellite, and personal communication systems. The first commercial analog cellular service was introduced in the U.S. in 1983. By the end of 1990, there were approximately 4.5 million cellular telephones in North America and about the same number in the rest of the world. The recent generation of cellular service uses time division multiple access (TDMA), code division multiple access (CDMA), narrow-band frequency division multiple access (FDMA), and collision sense multiple access (CSMA) spread spectrum.

The first public mobile telephone service was introduced in 20 major American cities in 1946. There are several incompatible standards employed in different parts of the world. Standardization of the global system for mobile communication (GSM) started in 1982 within the Conference of European Postal and Telecommunication Administrations (CEPT), the European spectrum management body. Since 1991, mobile radio has been dominated by the GSM system. The new generation of mobile/cellular systems will conform to at least three different standards: one for western Europe (GSM), one for North America, and one for Japan.

Paging is a form of organized mobile radio service. The first paging system was installed in 1956 in a London hospital. The first wide area paging systems were developed in the U.S. and Canada in the early 1960s. It was estimated in the early 1990s that there were more than 9 million pagers in the world.

Cellular mobile radio systems are becoming known in the U.S. as personal communications service (PCS). The emerging concepts of PCS and personal communications network (PCN) provide the freedom for users to communicate any type of information between any two points irrespective of where the users are physically located.

The first step of live global communication by earth satellite was made in 1962 when the Telstar 1 satellite was launched. By the mid-1980s, most TV stations received their network program feeds and much of their syndicated programming by satellite.

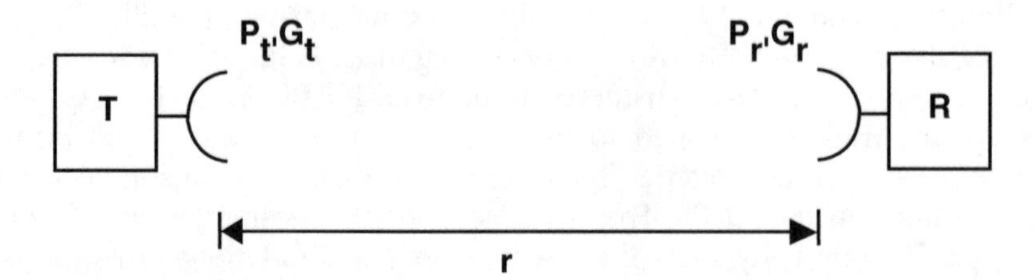

Figure 5.1 Basic wireless system.

5.2 Propagation characteristics

The major factors that affect the design and performance of wireless networks are the characteristics of radio or electromagnetic wave propagation over the geographical area. *Propagation* refers to the various ways by which an electromagnetic (EM) wave travels from the transmitting antenna to the receiving antenna. Propagation of EM waves can be regarded as a means of transferring energy or information from one point (a transmitter) to another (a receiver).

EM wave propagation can be described by two complementary models. The physicist attempts a theoretical model based on universal laws, which extends the field of application more widely than currently known. The engineer prefers an empirical model based on measurements, which can be used immediately. This chapter presents the complementary standpoints by discussing theoretical factors affecting wave propagation and the semi-empirical rules allowing handy engineering calculations. Here we describe the free space propagation model, path loss models, and the empirical path loss formula.

5.2.1 Free space propagation model

Wireless links typically experience free space propagation. The free space propagation model is used in predicting received signal strength when the transmitter and receiver have a clear line-of-sight path between them. If the receiving antenna is separated from the transmitting antenna in free space by distance r, as shown in Figure 5.1, the power received P_r by the receiving antenna is given by the Friis equation[3]

$$P_r = G_r G_t \left(\frac{\lambda}{4\pi r} \right)^2 P_t \tag{5.1}$$

where P_t is the transmitted power, G_r is the receiving antenna gain, G_t is the transmitting antenna gain, and λ is the wavelength ($= c/f$) of the transmitted signal.

The Friis equation relates the power received by one antenna to the power transmitted by the other, provided that the two antennas are separated by $r > 2d^2/\lambda$, where d is the largest dimension of either antenna. Thus, the Friis equation applies only when the two antennas are in the far-field of each other. It also shows that that the received power falls off as the square or the separation distance r. The power decay as $1/r^2$ in a wireless system, as exhibited in Equation 5.1, is better than the exponential decay in power in a wired link. In actual practice, the value of the received power given in Equation 5.1 should be taken as the maximum possible because some factors can reduce the received power in a real wireless system. This aspect is discussed fully in the next section.

From Equation 5.1, note that the received power depends on the product $P_t G_t$. The product is defined as the *effective isotropic radiated power* (EIRP), i.e.,

$$\text{EIRP} = P_t G_t \tag{5.2}$$

The EIRP represents the maximum radiated power available from a transmitter in the direction of maximum antenna gain relative to an isotropic antenna.

5.2.2 Path loss model

Wave propagation hardly occurs under the idealized conditions assumed in Section 5.2.1. For most communication links, the analysis in Section 5.2.1 must be modified to account for the presence of the earth, the ionosphere, and atmospheric precipitates such as fog, raindrops, snow, and hail.[4] This section considers these conditions.

The major regions of the earth's atmosphere of importance to radio wave propagation are the troposphere and ionosphere. At radar frequencies (approximately 100 MHz to 300 GHz), the troposphere is by far the most important. It is the lower atmosphere, a nonionized region extending from the earth's surface up to about 15 km. The ionosphere is the earth's upper atmosphere in the altitude region from 50 km to one earth radius (6370 km). Sufficient ionization exists in this region to influence wave propagation.

There are three modes of wave propagation over the surface of the earth:

- Surface wave propagation along the surface of the earth
- Space wave propagation through the lower atmosphere
- Sky wave propagation by reflection from the upper atmosphere.

These modes are portrayed in Figure 5.2. The sky wave is directed toward the ionosphere, which under certain conditions bends the propagation path back toward the earth in a limited frequency range (0–50 MHz approximately). This is highly dependent on the condition of the ionosphere (its level of ionization) and the signal frequency. The surface (or ground) wave takes effect at the low-frequency end of the spectrum (2–5 MHz approximately) and is directed along the surface over which the wave is propagated.

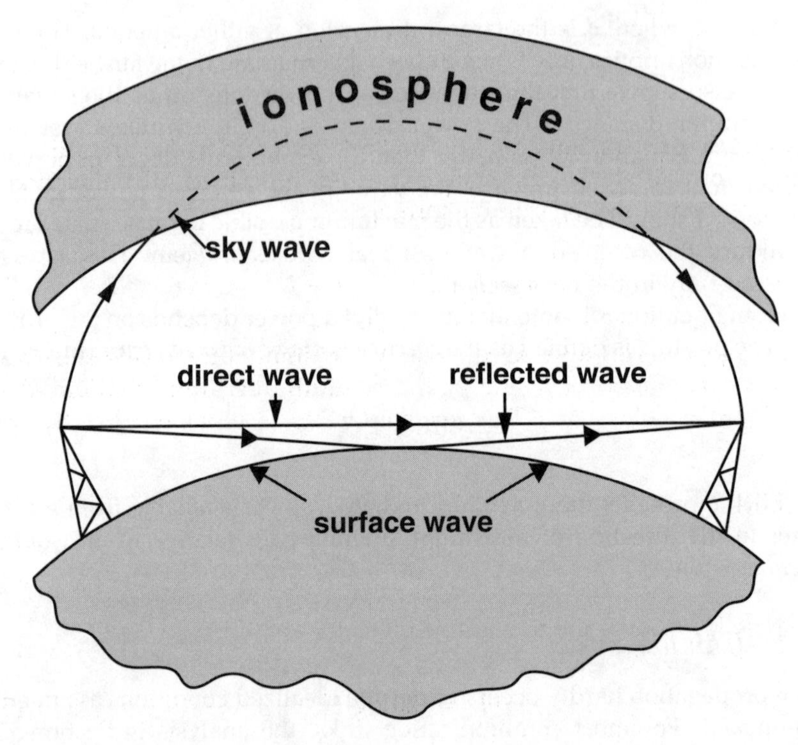

Figure 5.2 Modes of wave propagation.

Since the propagation of the ground wave depends on the conductivity of the earth's surface, the wave is attenuated more than if it were propagated through free space. The space wave consists of the direct wave and the reflected wave. The direct wave travels from the transmitter to the receiver in nearly a straight path while the reflected wave is due to ground reflection. The space wave obeys the optical laws in that direct and reflected wave components contribute to the total wave component. Although the sky and surface waves are important in many applications, this chapter considers only the space wave.

In case the propagation path is not in free space, a correction factor F is included in the Friis equation (5.1) to account for the effect of the medium. This factor, known as the *propagation factor*, is simply the ratio of the electric field intensity E_m in the medium to the electric field intensity E_o in free space, i.e.,

$$F = \frac{E_m}{E_o} \tag{5.3}$$

The magnitude of F is always less than unity since E_m is always less than E_o. Thus, for a lossy medium, Equation 5.1 becomes

$$P_r = G_r G_t \left(\frac{\lambda}{4\pi r} \right)^2 P_t |F|^2 \tag{5.4}$$

For practical reasons, Equations 5.1 and 5.4 are commonly expressed in logarithmic form. If all the terms are expressed in decibels (dB), Equation 5.4 can be written in the logarithmic form as

$$P_r = P_t + G_r + G_t - L_o - L_m \tag{5.5}$$

where P is power in dB referred to 1 W (or simply dBW), G is gain in dB, L_o is free-space loss in dB, and L_m is loss in dB due to the medium. (Note that G(dB) = $10\log_{10}$ G.) The free-space loss is obtained directly from Equation 5.4 as

$$L_o = 20\log\left(\frac{4\pi r}{\lambda} \right) \tag{5.6}$$

while the loss due to the medium is given by

$$L_m = -20\log|F| \tag{5.7}$$

The major concern of the rest of this subsection is to determine L_o and L_m for an important case of space propagation that differs considerably from the free-space conditions.

The phenomenon of multipath propagation causes significant departures from free-space conditions. The term *multipath* denotes the possibility of EM waves propagating along various paths from the transmitter to the receiver. In multipath propagation of an EM wave over the earth's surface, two such path exists: a direct path and a path via reflection and diffraction from the interface between the atmosphere and the earth. A simplified geometry of the multipath situation is shown in Figure 5.3. The reflected and diffracted component is commonly separated into two parts — one *specular* (or coherent) and the other *diffuse* (or incoherent) — which can be separately analyzed. The specular component is well defined in terms of its amplitude, phase, and incident direction. Its main characteristic is its conformance to Snell's law for reflection, which requires that the angles of incidence and reflection be equal and coplanar. It is a plane wave, and as such it is uniquely specified by its direction. The diffuse component, however, arises out of the random nature of the scattering surface and, as such, is nondeterministic. It is not a plane wave and does not obey Snell's law for reflection. It does not come from a given direction but from a continuum.

The loss factor F that accounts for the departures from free-space conditions is given by

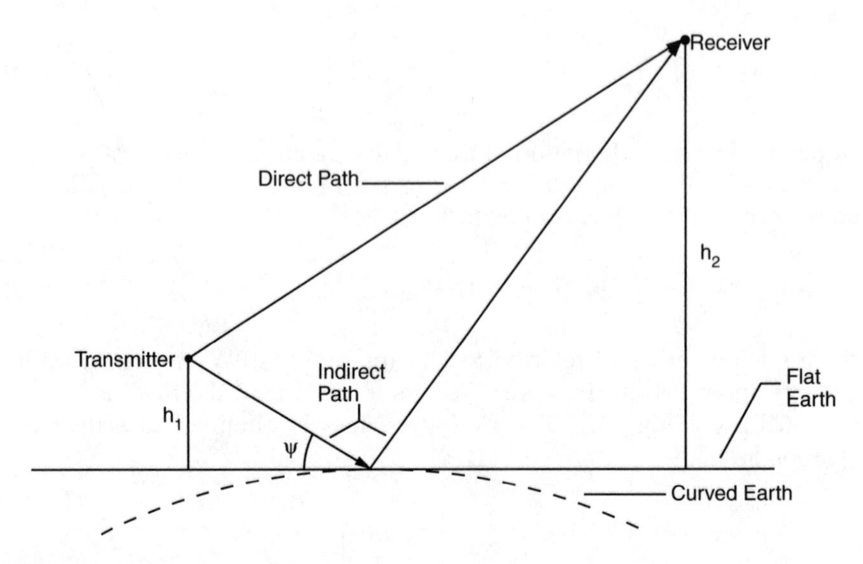

Figure 5.3 Multipath geometry.

$$F = 1 + \Gamma \rho_s D S(\theta) e^{-j\Delta} \tag{5.8}$$

where

Γ = Fresnel reflection coefficient
ρ_s = roughness coefficient
D = divergence factor
$S(\theta)$ = shadowing function
Δ = the phase angle corresponding to the path difference

The Fresnel reflection coefficient Γ accounts for the electrical properties of the earth's surface. Since the earth is a lossy medium, the value of the reflection coefficient depends on the complex relative permittivity ε_c of the surface, the grazing angle ψ, and the wave polarization. It is given by

$$\Gamma = \frac{\sin\psi - z}{\sin\psi + z} \tag{5.9}$$

where

$$z = \sqrt{\varepsilon_c - \cos^2\psi} \quad \text{for horizontal polarization} \tag{5.10}$$

$$z = \frac{\sqrt{\varepsilon_c - \cos^2\psi}}{\varepsilon_c} \quad \text{for vertical polarization} \tag{5.11}$$

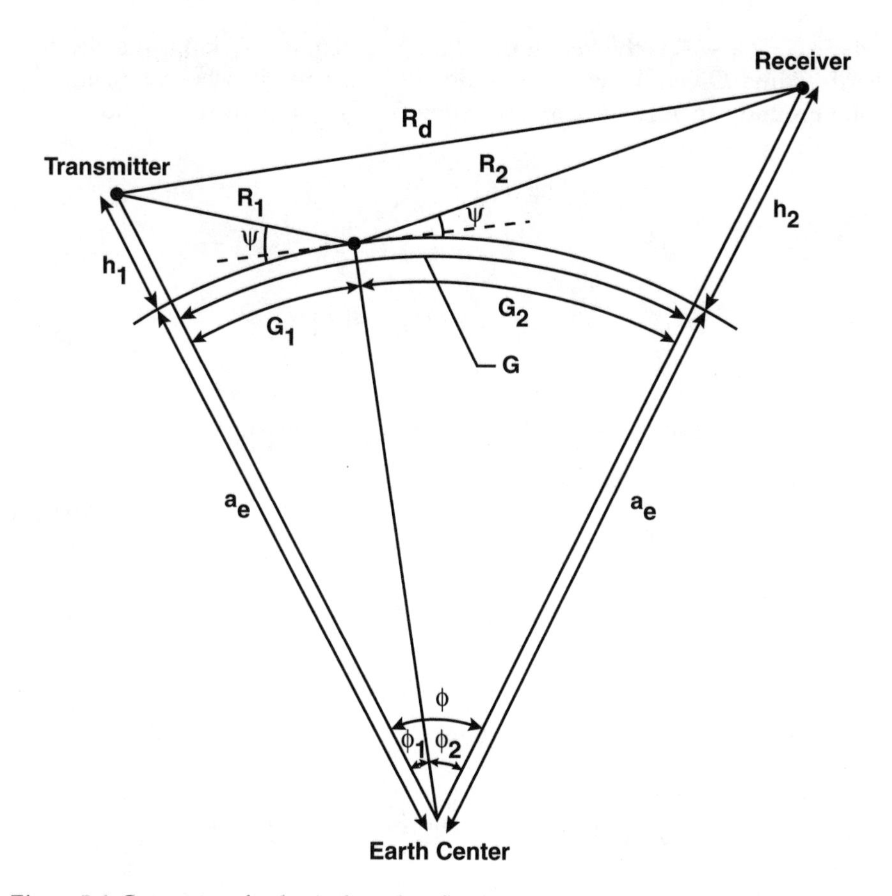

Figure 5.4 Geometry of spherical earth reflection.

$$\varepsilon_c = \varepsilon_r - j\frac{\sigma}{\omega\varepsilon_o} = \varepsilon_r - j60\sigma\lambda \tag{5.12}$$

ε_r and σ are, respectively, the dielectric constant and conductivity of the surface; ω and λ are, respectively, the frequency and wavelength of the incident wave; and ψ is the grazing angle. It is apparent that $0 < |\Gamma| < 1$.

To account for the spreading (or divergence) of the reflected rays due to the earth's curvature, the divergence factor D is introduced. The curvature has a tendency to spread out the reflected energy more than a corresponding flat surface does. The divergence factor is defined as the ratio of the reflected field from the curved surface to the reflected field from a flat surface. Using the geometry of Figure 5.4, D is given by

$$D = \left(1 + \frac{2G_1G_2}{a_eG\sin\psi}\right)^{-1/2} \tag{5.13}$$

where $G = G_1 + G_2$ is the total ground range and $a_e = 6370$ km is the effective earth radius. Given the transmitter height h_1, the receiver height h_2, and the total ground range G, we can determine G_1, G_2, and ψ. If we define

$$p = \frac{2}{\sqrt{3}}\left[a_e(h_1 + h_2) + \frac{G^2}{4}\right]^{1/2} \tag{5.14}$$

$$\alpha = \cos^{-1}\left[\frac{2a_e(h_1 - h_2)G}{p^3}\right] \tag{5.15}$$

and assume $h_1 \le h_2$, $G_1 \le G_2$, using small angle approximation yields[5]

$$G_1 = \frac{G}{2} + p\cos\left(\frac{\pi + \alpha}{3}\right) \tag{5.16}$$

$$G_2 = G - G_1 \tag{5.17}$$

$$\phi_i = \frac{G_i}{a_e}, \quad i = 1,2 \tag{5.18}$$

$$R_i = \left[h_i^2 + 4a_e(a_e + h_i)\sin^2(\phi_i/2)\right]^{1/2}, \quad i = 1,2 \tag{5.19}$$

The grazing angle is given by

$$\psi = \sin^{-1}\left[\frac{2a_e h_1 + h_1^2 - R_1^2}{2a_e R_1}\right] \tag{5.20a}$$

or

$$\psi = \sin^{-1}\left[\frac{2a_e h_1 + h_1^2 + R_1^2}{2(a_e + h_1)R_1}\right] - \phi_1 \tag{5.20b}$$

Although D varies from 0 to 1, in practice D is a significant factor at low grazing angle ψ (less than 0.1%).

The phase angle corresponding to the path difference between direct and reflected waves is given by

$$\Delta = \frac{2\pi}{\lambda}(R_1 + R_2 - R_d) \tag{5.21}$$

The roughness coefficient ρ_s takes care of the fact that the earth's surface is not sufficiently smooth to produce a specular (mirror-like) reflection except at a very low grazing angle. The earth's surface has a height distribution that is random in nature. This randomness arises out of the hills, structures, vegetation, and ocean waves. It is found that the distribution of the heights of the earth's surface is usually the Gaussian or normal distribution of probability theory. If σ_h is the standard deviation of the normal distribution of heights, we define the roughness parameters

$$g = \frac{\sigma_h \sin\psi}{\lambda} \tag{5.22}$$

If $g < 1/8$, specular reflection is dominant; if $g > 1/8$, diffuse scattering results. This criterion, known as the *Rayleigh criterion*, should be used only as a guideline since the dividing line between a specular and a diffuse reflection or between a smooth and a rough surface is not well defined.[6] The roughness is taken into account by the roughness coefficient ($0 < \rho_s < 1$) — the ratio of field strength after reflection with roughness taken into account to that which would be received if the surface were smooth. The roughness coefficient is given by

$$\rho_s = \exp\left[-2(2\pi g)^2\right] \tag{5.23}$$

Shadowing is the blocking of the direct wave by obstacles. The shadowing function $S(\theta)$ is important at a low grazing angle. It considers the effect of geometric shadowing — the fact that the incident wave cannot illuminate parts of the earth's surface shadowed by higher parts. In a geometric approach, where diffraction and multiple scattering effects are neglected, the reflecting surface will consist of well-defined zones of illumination and shadow. As there will be no field on a shadowed portion of the surface, the analysis should include only the illuminated portions of the surface. A pictorial representation of rough surfaces illuminated at angle of incidence $\theta(= 90° - \psi)$ is shown in Figure 5.5. It is evident from the figure that the

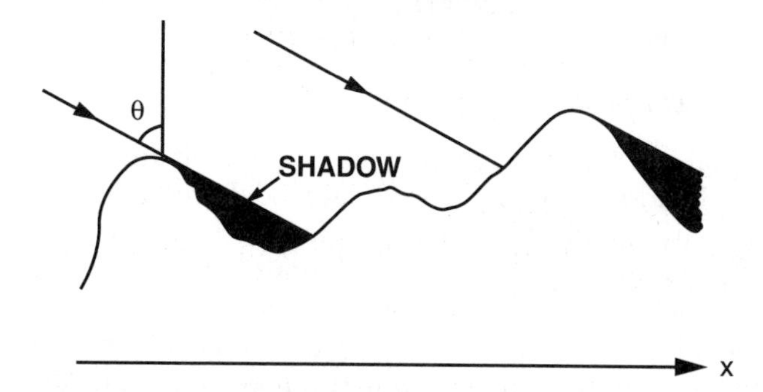

Figure 5.5 Rough surface illuminated at an angle of incidence θ.

shadowing function $S(\theta)$ is equal to unity when $\theta = 0$ and zero when $\theta = \pi/2$. According to Smith[7]

$$S(\theta) = \frac{\left[1 - \frac{1}{2}\text{erfc}(a)\right]}{1 + 2B} \tag{5.24}$$

where erfc(x) is the complementary error function,

$$\text{erfc}(x) = 1 - \text{erfc}(x) = \frac{2}{\sqrt{\pi}} \int_x^\infty e^{-t^2} dt \tag{5.25}$$

and

$$B = \frac{1}{4a}\left[\frac{1}{\sqrt{\pi}} e^{a^2} - a\,\text{erfc}(a)\right] \tag{5.26}$$

$$a = \frac{\cot\theta}{2s} \tag{5.27}$$

$$s = \frac{\sigma_h}{\sigma_l} = \text{rms surface slope} \tag{5.28}$$

In Equation 5.28, σ_η is the rms roughness height and σ_l is the correlation length. Alternative models for $S(\theta)$ are available in the literature. Using Equations 5.9 through 5.28, the loss factor in Equation 5.8 can be calculated. Thus

$$L_o = 20\log\left(\frac{4\pi R_d}{\lambda}\right) \tag{5.29}$$

$$L_m = -20\log\left[1 + \Gamma\,\rho_s\, D\, S(\theta)\, e^{-j\Delta}\right] \tag{5.30}$$

5.2.3 *Empirical path loss formula*

Both theoretical and experimental propagation models are used in predicting the path loss. In addition to the theoretical model presented in the previous section, there are empirical models for finding path loss. Of the several models in the literature, the model of Okumura et al.[8] is the most popular for analyzing mobile-radio propagation because of its simplicity and accuracy. The model is based on extensive measurements taken in and around Tokyo and then compiled into charts that can be applied to VHF and UHF mobile-radio propagation. The medium path loss is given by[9]

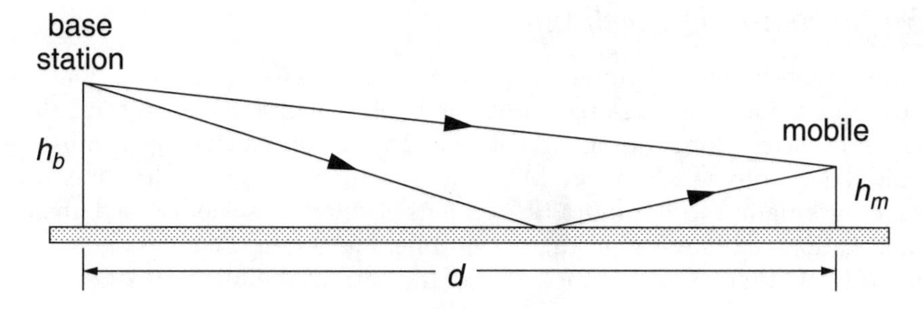

Figure 5.6 Radio propagation over a flat surface.

$$L_p = \begin{cases} A + B\log_{10}(r) & \text{for urban area} \\ A + B\log_{10}(r) - C & \text{for suburban area} \\ A + B\log_{10}(r) - D & \text{for open area} \end{cases} \tag{5.31}$$

where r (in kilometers) is the distance between the base and mobile stations, as illustrated in Figure 5.6. The values of A, B, C, and D are given in terms of the carrier frequency f, the base station antenna height h_b (in meters), and the mobile station antenna height h_m (in meters) as

$$A = 69.55 + 26.16\ \log_{10}(f) - 13.82\ \log_{10}(h_b) - \mathrm{a}(h_m) \tag{5.32a}$$

$$B = 44.9 - 6.55\ \log_{10}(h_b) \tag{5.32b}$$

$$C = 5.4 + 2\left[\log_{10}\left(\frac{f}{28}\right)\right]^2 \tag{5.32c}$$

$$D = 40.94 - 19.33\log_{10}(f) + 4.78\left[\log_{10}(f)\right]^2 \tag{5.32d}$$

where

$$a(h_m) = \begin{cases} 0.8 - 1.56\log_{10}(f) + \left[1.1\log_{10}(f) - 0.7\right]h_m, & \text{for medium/small city} \\ 8.28\left[\log_{10}(1.54h_m)\right]^2 - 1.1, & \text{for } f \geq 200 \text{ MHz} \\ 3.2\left[\log_{10}(11.75h_m)\right]^2 - 4.97, & \text{for } f < 400 \text{ MHz} \\ & \text{for large city} \end{cases} \tag{5.33}$$

The following conditions must be satisfied before Equation 5.32 is used: $150 < f < 1500$ MHz; $1 < r < 80$ km, $30 < h_b < 400$ m; $1 < h_m < 10$m. Okumura's model has been found to be fairly good in urban and suburban areas, but not as good in rural areas.

5.3 Modulation techniques

Information is often represented in some form of electrical pulses that are not suitable for transmission over the air. In other words, baseband signaling is not sufficient for sending signals over any distance. The signal must be modulated onto an RF carrier. Modulation is the means of adding the message information to the radio carrier. Earlier wireless systems used analog modulation techniques, but present and next generation wireless networks use digital modulation schemes. Two of the various digital modulation techniques widely used for wireless systems are $\pi/4$ phase-shifted quadrature phase shift keying ($\pi/4$-QPSK) in North America and Japan and Gaussian minimum shift keying (GMSK) in the European GSM system.

A sinusoidal carrier may be represented as

$$s(t) = A(t)\cos(\omega t + \varphi) \tag{5.34}$$

where A is the amplitude, $\omega = 2\pi f$ is the angular frequency, and φ is the phase. Depending on which of the three quantities is modified or modulated, there are three kinds of analog modulation techniques: amplitude modulation (AM), frequency modulation (FM), and phase modulation (PM). In a similar manner, there are three kinds of modulation techniques for digital signals:

- Amplitude shift keying (ASK), when ω and φ are unchanged
- Frequency shift keying (FSK), when A and φ are unchanged
- Phase shift keying (PSK), when A and ω remain unchanged

The most common method is to fix ω and change A and φ. This is known as *quadrature amplitude modulation* (QAM).

In the quadrature phase shift keying (QPSK) modulation technique, the amplitude is fixed while four different phase angles are used leading to four symbols: $\pi/4$, $3\pi/4$, $5\pi/4$, and $7\pi/4$. The modulated carriers with the four phase angles and constant amplitude are expressed as phasors or vectors as shown in Figure 5.7. The QPSK modulated carrier can be written as[10]

$$s_0(t) = A\cos(\omega t + 45°) \tag{5.35a}$$

$$s_1(t) = A\cos(\omega t + 135°) \tag{5.35b}$$

$$s_2(t) = A\cos(\omega t - 135°) \tag{5.35c}$$

$$s_3(t) = A\cos(\omega t - 45°) \tag{5.35d}$$

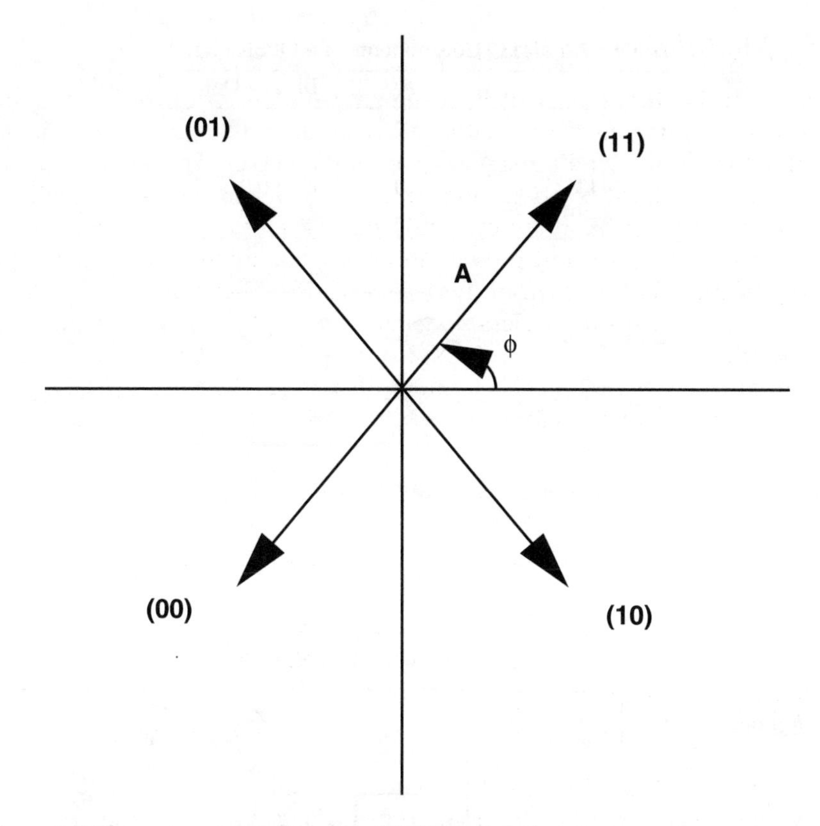

Figure 5.7 Phasor representation of a QPSK modulation.

or in a more compact way as

$$s_i(t) = A\cos(\omega t + \phi_i), \quad \phi_i = (2i+1)\pi/4, \quad i = 0,1,2,3 \tag{5.36}$$

This can be expanded in terms of the in-phase (cosine) and quadrature (shifted 90° in phase) components using trigonometric identity as

$$s_i(t) = A_I \cos\omega t + A_Q \sin\omega t \tag{5.37}$$

where A_I and A_Q are provided in Table 5.1, where the common factor of $\sqrt{2}/2$ has been ignored. Equation 5.37 suggests that the QPSK signal can be generated by using binary NRZ data to modulate the in-phase (cosine) and quadrature (sine) components of the carrier, as illustrated in Figure 5.8. The symbol-to-bit mapping is such that as the phase angle changes from the neighboring angle, the two-bit pattern makes one bit change. This type of symbol-to-bit mapping is known as Gray coding. The advantage is that a single symbol error corresponds to single bit error. Also, since each symbol

Table 5.1 I and Q Components of a QPSK Signal

i	ϕ_i	A_I	A_Q	Binary Data
0	45°	1	1	1,1
1	135°	–1	1	0,1
2	–135°	–1	–1	0,0
3	–45°	1	–1	1,0

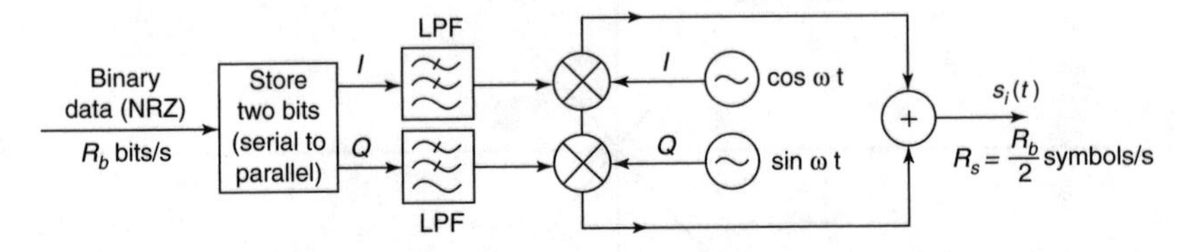

Figure 5.8 A QPSK modulator.

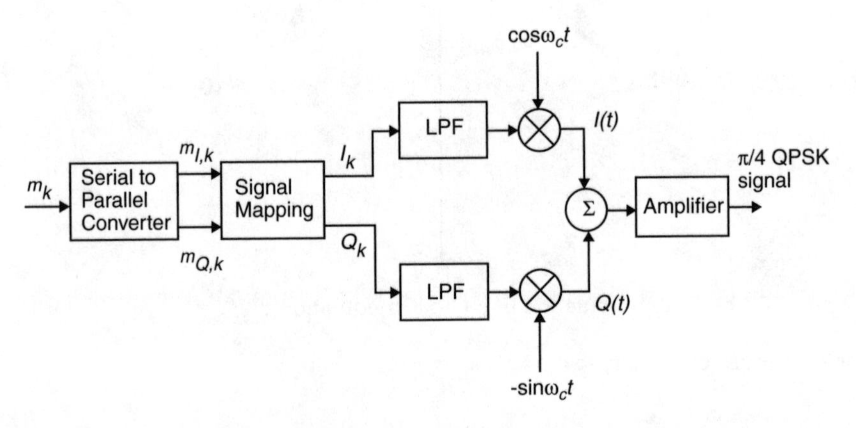

Figure 5.9 A π/4-QPSK modulator.

(one of four phases) represents two bits, the bit transmission rate is twice the symbol rate.

Another modulation scheme, π/4-QPSK, belongs to the class of QPSK. Unlike the conventional QPSK, which has four phases, π/4-QPSK has eight possible transmitted phases. In π/4-QPSK, the maximum phase change is limited to ±135° compared with 180° for QPSK. Hence, the π/4-QPSK signal preserves the constant envelope property better than does QPSK. A block diagram of a π/4-QPSK modulator is presented in Figure 5.9. An attractive property of π/4-QPSK is that it can be noncoherently detected, which simplifies receiver design. When differentially encoded, π/4-QPSK is known as π/4-DQPSK.

The performance of a modulation technique is often determined by its bandwidth efficiency. Bandwidth efficiency η_B is the ability of a modulation method to accommodate data within a limited bandwidth. If B is the bandwidth (in Hz) occupied by the modulated signal and R is the data rate (in bps), then

$$\eta_B = \frac{R}{B}(\text{bps/Hz}) \tag{5.38}$$

The capacity of a digital communication system is related to the bandwidth efficiency. There is an upper limit to an achievable bandwidth efficiency. According to Shannon's channel theorem, the maximum possible bandwidth efficiency is limited by noise in the channel and is given by

$$\eta_{B\max} = \frac{C}{B} = \log_2\left(1 + \frac{S}{N}\right) \tag{5.39}$$

where C is the channel capacity (in bps), B is the bandwidth (in Hz), and S/N is the signal-to-noise ratio. Equation 5.39, known as the *Shannon channel capacity formula*, gives the upper bound on the maximum data rate achievable for a given channel in the presence of noise.

5.4 Multiple access techniques

Multiple access schemes were discussed in Chapter 3 but are briefly discussed here in relation to wireless communication systems. Figure 5.10 illustrates the complete frequency spectrum and where different types of waves fit.

Spectrum is a scarce and limited resource, so multiple access schemes are designed to share the resource among a large number of wireless users.

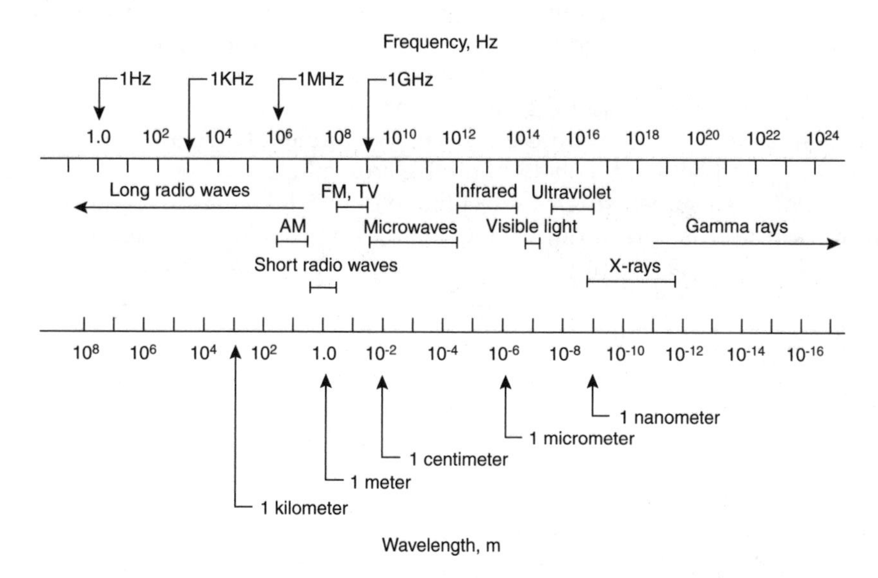

Figure 5.10 The electromagnetic spectrum.

There are three popular multiple access techniques for sharing the available bandwidth in a wireless communication system:

- Frequency division multiple access (FDMA) serves users with different frequency channels. Signals are transmitted in nonoverlapping frequency bands that can be separated using bandpass filters.
- Time division multiple access (TDMA) serves users with different time slots. Signals are transmitted in nonoverlapping time slots in a round-robin fashion. In each slot only one user is allowed either to transmit or receive.
- Code division multiple access (CDMA) serves users with different code sequences. Different users employ signals that have small cross-correlation.

The three access methods are illustrated in Figure 5.11. In addition to FDMA, TDMA, and CDMA, there are two other multiple access schemes — polarization division multiple access (PDMA) which serves users with different polarization and space division multiple access (SDMA) which controls the radiated energy for users in space by using a spot beam antenna.[11]

FDMA assigns individual channels to individual users. As shown in Figure 5.11, guard bands are maintained between adjacent spectra to minimize crosstalk between channels. Present FM radio subdivides the spectrum into 30-kHz channels so that each channel is assigned to one user. In FDMA, the 30-kHz channel can be split into three 10-kHz channels. This bandsplitting, however, incurs costs. TDMA and CDMA can support more users in the space spectrum region.

TDMA is a channelization scheme that triples the capacity of the available channels without requiring additional RF spectrum. A frame consists of a number of time intervals called slots. As shown in Figure 5.12, each TDMA frame consists of a preamble, information message, and trail bits. The preamble has the address and synchronization information that both the base station and the subscribers will use to identify each other. Guard times are used between slots to minimize crosstalk.

CDMA is a spread spectrum technique in which the narrowband signal from each user is spread out in frequency using a unique spreading code. Several signals may occupy the same frequency band and still be individually recovered by the receiver with the knowledge of the spreading code. Each user operates independently of other users.

Summary

Wireless communication is growing quickly. Wireless networks include wireless LANs, mobile field services, point-to-point connections between remote stations without wires, and WAN connections without lease line services. This chapter discussed some fundamental concepts in wireless communication systems, including wave propagation with different models, modulation

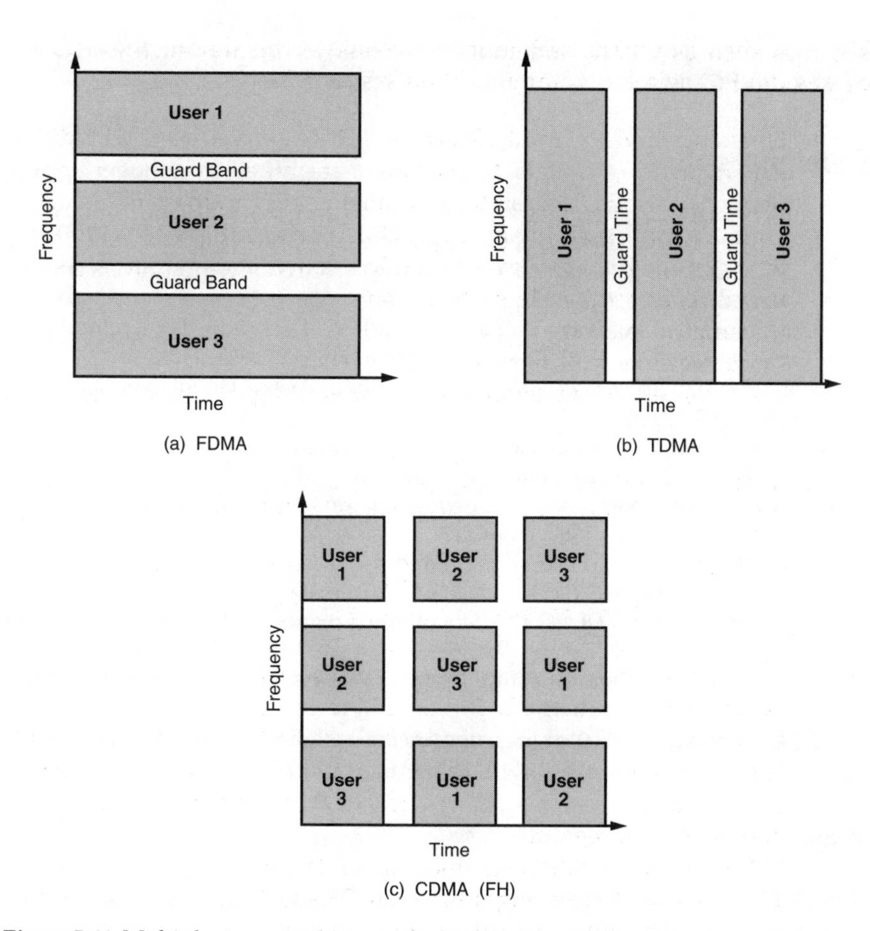

Figure 5.11 Multiple access techniques for wireless systems.

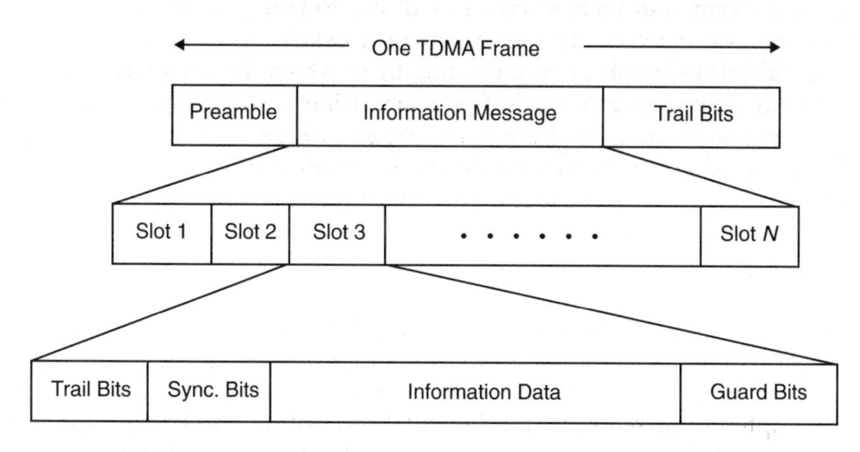

Figure 5.12 TDMA frame structure.

schemes such as QPSK, and multiple access techniques such as TDMA, FDMA, and CDMA.

References

1. M. Nemzow, *Implementing Wireless Networks*, McGraw-Hill, New York, 1995, 3.
2. B. Bates, *Wireless Networked Communications*, McGraw-Hill, New York, 1993, 1.
3. M. N. O. Sadiku, *Elements of Electromagnetics*, 3rd ed., Oxford University Press, New York, 2001, 621–623.
4. M. N. O. Sadiku, Wave Propagation, in R. C. Dorf (Ed.), *The Electrical Engineering Handbook*, CRC Press, Boca Raton, FL, 1997, 925–937.
5. L. V. Blake, *Radar Range-Performance Analysis*, Artech House, Norwood, MA, 1986, 253–271.
6. P. Beckman and A. Spizzichino, *The Scattering of Electromagnetic Waves from Random Surfaces*, Macmillan, New York, 1963.
7. B. G. Smith, Geometrical shadowing of a random rough surface, *IEEE Trans. Ant. Prog.*, vol. 15, 1967, 668–671.
8. Y. Okumura et al., Field strength and its variability in VHF and UHF land mobile service, *Rev. Electr. Comm. Lab*, vol. 16, Sept./Oct. 1969, 825–873.
9. K. Feher, *Wireless Digital Communications*, Prentice-Hall, Upper Saddle River, NJ, 1995, 74–76.
10. D. M. Pozar, *Microwave and RF Design of Wireless Systems*, John Wiley & Sons, New York, 2001, 324–327.
11. T. S. Rappaport, *Wireless Communications: Principles and Practice*, Prentice-Hall, Upper Saddle River, NJ, 1996, 395–438.

Problems

5.1 Discuss two current wired services that will migrate to a wireless environment. Give reasons for your answers.

5.2 Right now there are several standards for mobile/cellular systems. Explain why there should be one worldwide standard.

5.3 The transmitting and receiving antennas are 10 km apart and have directive gains of 20 and 25 dB, respectively. If the required received power is 4 nW, what is the minimum transmitted power required for transmission at 3 GHz? (Note that G (dB) = 10 $\log_{10}$ G.)

5.4 Show that the Brewster angle (when $\Gamma = 0$ for vertical polarization) is given by ψ where

$$\sin\psi = \frac{1}{\sqrt{\varepsilon_r + 1}}$$

when $\sigma = 0$.

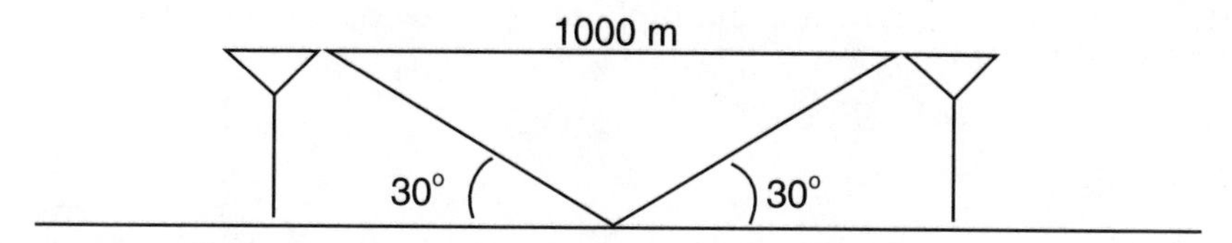

Figure 5.13 For Problem 5.5.

5.5 Consider the direct and reflected paths shown in Figure 5.13. Calculate the phase difference at 25 MHz.

5.6 Describe at least three causes of medium loss in wave propagation.

5.7 Apply the Okumura model to find the path loss in a suburban area over a distance of 12 km at a frequency of 600 MHz. Assume that the base-station antenna height is 60 m and the mobile antenna height is 4 m.

5.8 Explain the difference between QPSK and $\pi/4$-QPSK.

5.9 Determine the theoretical maximum data rate that can be supported in a wireless communication link with signal-to-noise ratio of 30 dB and RF bandwidth of 40 kHz.

5.10 What is the S/N requirement of a 2.5 Mbps modem if the bandwidth equals 2 MHz?

chapter six

Wireless networking

Good fortune is what happens when opportunity meets with planning.

— **Thomas Edison**

Having covered the fundamentals common to wireless communications systems, this chapter and the following two chapters consider some wireless communication systems, including:

- Wireless local area networks (WLANs), which enable communication between stations without cables by means of radio frequency or infrared
- Wireless asynchronous transfer mode (WATM), which facilitates the use of ATM technology for a broad range of wireless network access and internetworking scenarios
- Wireless local loop (WLL) or fixed radio, which provides telephone, fax, and data services
- Wireless private branch exchanges (WPBXs), which facilitate communication with the office environment, allowing workers to roam
- Wireless personal area networks (WPAN), which use a near-field electric field to send data across various devices using the human body as a medium
- Wireless personal communications services (PCS), all access technologies used by individuals or subscribers
- Cellular communications, which allows frequency reuse by dividing regions into small cells, each cell with a stationary radio antenna
- Satellite communications, which uses orbiting satellites to relay data between multiple earth-based stations

The first five systems are considered in this chapter; the others are discussed in Chapters 7 and 8.

6.1 Wireless LAN

Wireless local area network (WLAN) is a new form of communication system. It is basically a local area network, confined to a geographically small area, such as a single building, office, store, or campus, that provides high data connectivity to mobile stations. Using electromagnetic airwaves (radio frequency or infrared), WLANs transmit and receive data over the air. A WLAN suggests less expensive, fast, and simple network installation and reconfiguration.

The proliferation of portable computers coupled with the mobile user's need for communication is the major driving force behind WLAN technology. WLAN creates a mobile environment for the PC and LAN user. It may lower LAN maintenance and expansion costs since there are no wires that require reconfiguration. Thus, WLANs offer the following advantages over conventional wired LANs:

- Installation flexibility — the network can go where wire cannot
- Mobility — LAN users can have access anywhere
- Scalability — it can be configured in a variety of topologies to meet specific needs

However, WLAN does not perform as well as wired LAN because of the bandwidth limitations, and it can be susceptible to electromagnetic interference and distance limitations. Although the initial investment in WLAN hardware can be higher than the cost of wired LAN hardware, overall installation expenses and life-cycle costs can be significantly lower.

6.1.1 Physical layer and topology

WLAN does not compete with wired LAN. Rather, WLANs are used to extend wired LANs for convenience and mobility; they essentially "fill" in for wired links using electromagnetic radiation at radio or light frequencies between transceivers. A typical WLAN consists of an access point and the WLAN adapter installed on a portable notebook. The access point is a transmitter/receiver (transceiver) device; it is the wireless equivalent of a regular LAN hub. An access point is typically connected with the wired backbone network at a fixed location through a standard Ethernet cable, and it communicates with wireless devices by antenna. WLANs operate within the prescribed 900 MHz, 2.4 GHz, and 5.8 GHz frequency bands. Most LANs use 2.4 GHz frequency bands because it is most widely accepted.

A wireless link can provide services in several ways, including the following three:[1]

- *Replace a point-to-point connection between two nodes or segments on a LAN:* A point-to-point link is a connection between two devices for transferring data. A wireless link can be used to bridge two LAN segments, as shown in Figure 6.1. Like a point-to-point link, the link connects two wireless bridges attached to the two LANs. Such an

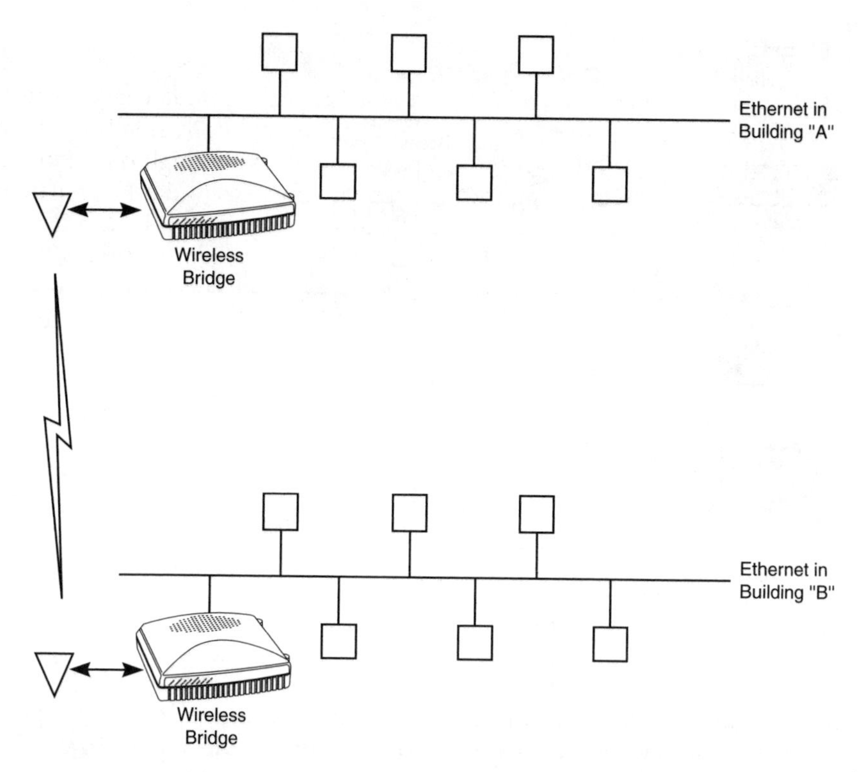

Figure 6.1 A wireless link replacing a point-to-point connection.

arrangement is useful for linking LANs in two buildings where a highway or river makes direct connection difficult.

- *Provide a connection between a wired LAN and one or more WLAN nodes:* In this case, a device is attached to the wired LAN to act as a point of contact (called *access point*) between the wired LAN and the wireless nodes, as shown in Figure 6.2. The device can be a repeater, bridge, or router.
- *Act as a stand-alone WLAN for a group of wireless nodes:* This can be achieved using topologies similar to wired LAN; a star topology can be formed with a central hub controlling the wireless nodes, a ring topology with each wireless node receiving or passing information sent to it, or a bus topology with each wireless capable of hearing everything said by all the other nodes. These three popular WLAN topologies (star, ring, and bus) are shown in Figure 6.3.

6.1.2 Technologies

When designing WLANs, manufacturers must choose between two main technologies that are used for wireless communications today: radio frequency (RF) and infrared (IR). Each technology has its own merits and disadvantages.

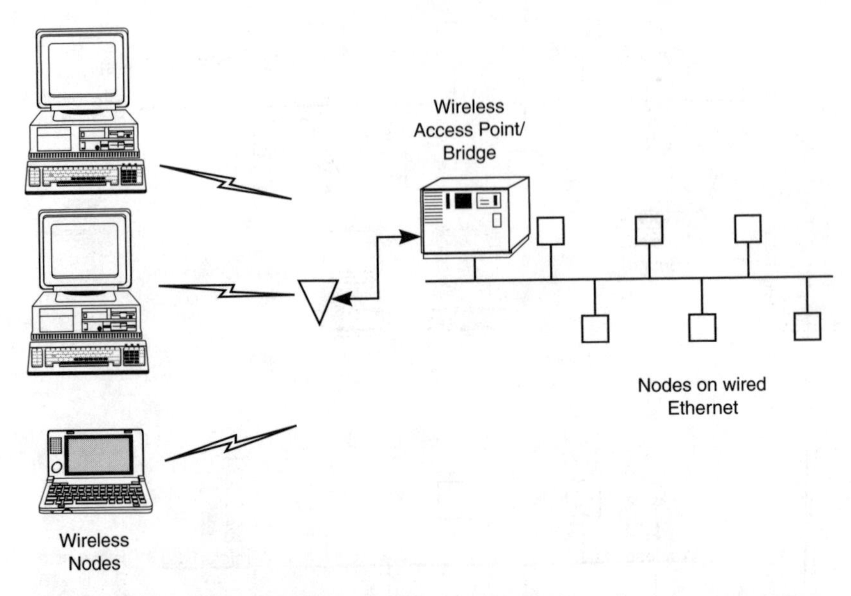

Figure 6.2 Connection of a wired LAN to wireless nodes.

RF is used for communications over long distances that are not line-of-sight. In order to operate in the license-free portion of the frequency spectrum, known as the ISM band (industrial, scientific, and medical), the RF system must use a modulation technique called *spread spectrum* (SS). Spread spectrum is wideband radio frequency technology developed by the military during World War II for use in reliable, secure, mission-critical communications systems. An SS system is one in which a transmitted signal is spread over a frequency much wider than the minimum bandwidth required to send the signal. Using spread spectrum, a radio is supposed to distribute the signal across the entire spectrum. Therefore, no single user can dominate the band, and collectively all users look like noise. The fact that such signals appear like noise in the band makes them difficult to find and jam, thereby increasing security against unauthorized listeners. There are two types of spread spectrum technology: frequency hopping and direct sequence.

Currently, frequency hopping spread spectrum (FHSS) offers a maximum data rate of 3 Mbps. It uses a narrowband carrier that changes frequency in a pattern known to both transmitter and receiver. It is based on the use of a signal at a given frequency that is constant for a small amount of time and then moves to a new frequency. The sequence of different channels for the hopping pattern is determined pseudorandomly — a very long sequence code is used before the sequence is repeated, over 65,000 hops, making it appear random. Thus, it is very difficult to predict the next frequency at which such a system will transmit/receive data as the system appears to be a noise source to an unauthorized listener, which makes FHSS very secure against interference and interception. FHSS is characterized by

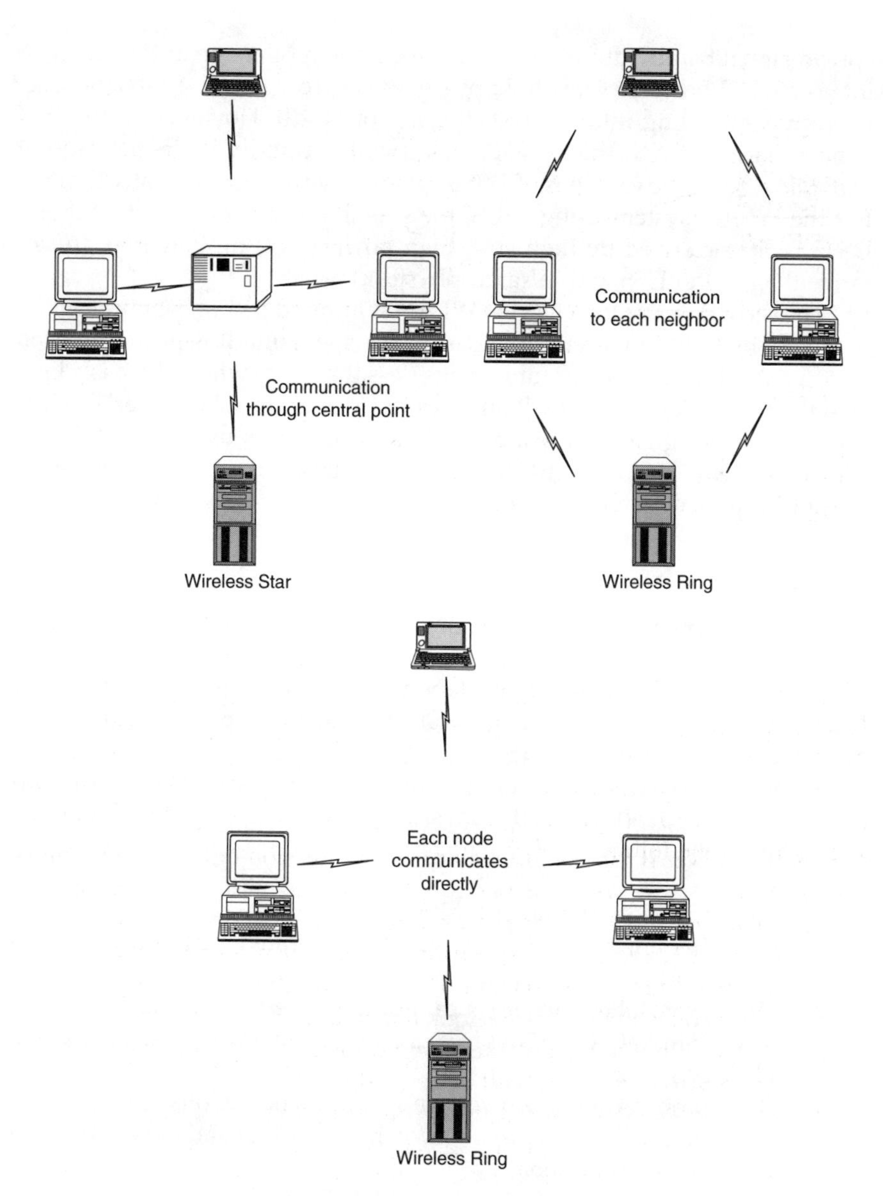

Figure 6.3 Stand-alone WLAN configurations: star, ring, and bus topologies.

low cost, low power consumption, and less range than DSSS but greater range than infrared. Most WLAN systems use FHSS.

Direct sequence spread spectrum takes a signal at a given frequency and spreads it across a band of frequencies where the center frequency is the original signal. The spreading algorithm, which is the key to the relationship of the spread range of frequencies, changes with time in a pseudorandom sequence. When the ratio between the original signal bandwidth and the spread signal bandwidth is very large, the system offers great immunity to interference. For example, if 10 kbps signal is spread across 1 GHz of spectrum, the spreading ratio is 100,000 times or 50 dB. However, in the ISM band used in WLAN, the available bandwidth critically limits the ratio of spreading, so the advantages of DSSS against interference is greatly limited. For the WLAN system using DSSS, the spreading ratio is at best 10 times. DSSS is characterized by high cost, high power consumption, and greater range than with FHSS and infrared physical layers.

The second technology used in WLAN is infrared (IR) — communication is carried by light in the invisible part of the spectrum. It is primarily used for very short distance communications (less than 1 m) where there is a line-of-sight connection. Since IR light does not penetrate solid materials (it is even attenuated greatly by window glass), it is not as useful as RF in the WLAN system. However, IR is used in applications in which power is extremely limited, such as pagers.

6.1.3 Standards

Although a number of proprietary, nonstandard wireless LANs exist, standards have now been developed. Two international organizations have contributed to the development of standards for WLANs: the Institute of Electronics and Electrical Engineers (IEEE) and the European Telecommunications Standards Institute (ETSI).

In 1997, the IEEE 802.11 committee (http://ieee802.org/11) issued a standard for WLANs. It addresses the physical and MAC layers of the OSI model and includes the following:[1,2]

- A transmission rate up to 2 Mbps.
- Two different media for transmission: infrared (IR) and radio frequency (RF).
- MAC protocol as carrier sense multiple access with collision avoidance (CSMA/CA), i.e., devices can interoperate with wired LANs via a bridge.
- MAC protocol with two service types: asynchronous and synchronous (or contention-free) — asynchronous is mandatory while the synchronous is optional.
- MAC layer protocol tied to the IEEE 802.2 logical link control (LLC) layer, making it easier to integrate with other LANs.
- Three different physical layers: an optical-based physical-layer implementation that uses IR light to transmit, two RF-based physical-layer choices: DSSS and frequency hopping FHSS, both operating at 2.4 GHz industrial, scientific, and medical (ISM) frequency bands. (The ISM bands 902–928 MHz, 2400–2483.5 MHz, and 5725–5850 MHz do not require a license to operate.) The IEEE 802.11 specifications for DSSS wireless LAN is shown in Figure 6.4.

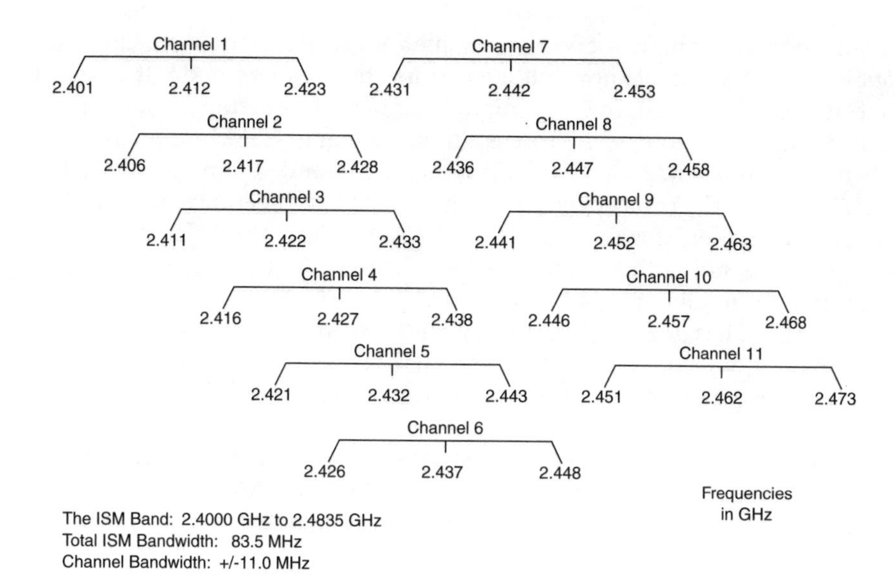

Figure 6.4 Eleven 22-MHz-wide channels for DSSS wireless LANs.

- Added features to the MAC that can maximize battery life in portable clients via power-management schemes.
- Data security through which the wireless LANs can achieve wired-equivalent privacy.

The standard defines the media and configuration issues, transmission procedures, throughput requirements, and range characteristics for WLAN technology. It avoids rigid requirements and gives room for vendors in the following areas: multiple physical media, common MAC layer irrespective of the physical layer, common frame format, power limit, and multiple on-air data rates[3].

There are three major problems encountered by an RF LAN.[4] First, frequency allocation is limited for LANs. But since LANs operate with low power, frequency reuse is possible. Second, interference from other WLANs controlled by a different organization and from other wireless sources is a problem. This problem can be controlled by using spread spectrum techniques. Third, security is at stake because the RF signal can penetrate through walls, and hostile operators can intercept RF LAN communications. Encryption can be used to lessen this problem. IR LAN uses both laser diodes and light-emitting diodes as emitters and is useful in high electromagnetic interference (EMI) environments. It is also secure since IR signals cannot penetrate the wall.

CSMA/CA is slightly different from carrier sense multiple access with collision detection (CSMA/CD), which is the MAC protocol used in Ethernet wired LAN. In CSMA/CA, when a node has something to transmit, it waits for silence on the network. When no other nodes are heard, it transmits and waits to receive an acknowledgment from the recipient node. If it fails to receive an acknowledgment within a time period, it assumes that collision has occurred and follows a process similar to that of CSMA/CD. Each node then waits for silence and transmits only after a random amount of waiting. While CSMA/CA protocol is slower than CSMA/CD because of waiting for acknowledgment, it works well for WLANs. Also, WLANs operate in a strong multipath fading channel where channel characteristics can change, resulting in unreliable communication.

The ETSI has devoted its attention to RF WLANs and is close to finalizing its standard, which is based on the 2.4 GHz range used for spread-spectrum LANs in several European countries. European standard WLAN, called HiperLAN, will allow speeds of 24 Mbps.[5]

Some organizations focus on the implementation and interoperability of WLAN products. Such organizations include Wireless LAN Alliance (WLANA at www.wlana.com) and Wireless Ethernet Compatibility Alliance (WECA at www.wi-fi.org or www.wirelessethernet.com). WLANA was formed in 1996 with 12 members as a trade association for WLAN vendors. WECA is a nonprofit manufacturing consortium with over 60 companies as members; it was formed in 1999 to certify interoperability of IEEE 802.11 products.

6.1.4 Applications

Offering the obvious advantage of no wire-installation costs, WLANs can be deployed in dynamic environments where there is frequent reconfiguration of computer networks. Also, without cables, excavation, or long installation time, it is simpler to connect difficult-to-reach customers.

Although several products for RF and IR LANs are already available in the marketplace, the introduction of their applications is just beginning. Typical mobile uses are laptop or notebook computers and portable base stations. Services provided by WLANs include data applications over TCP/IP and multimedia applications.

The most prominent users of WLANs are those whose projects promise quick payoffs for adding mobility. Industries such as security services, banks, retail, manufacturing, and health care are notable for deploying WLANs that allow workers to roam while gathering information.

Mobile terminals — personal digital assistants (PDAs), specialized handheld terminals, and barcode scanners — connected to WLANs are increasingly used to enhance business operations. It has become commonplace for WLANs to be used in the applications:[1]

- Printer sharing: linking to a distant printer within a department
- Electronic mail: sending and receiving e-mails from anywhere
- Health care: access to patient records from practically anywhere and location-independent claims processing

- Financial services such as Stock or Community Exchange: implementing hand-held communicators in the trading room to increase the speed, accuracy, and reliability of its price reporting system
- Factory control: data acquisition, inventory control, scoreboards, and robotics

Other applications include trading, banking, restaurants, retail, warehousing, manufacturing, education, offices, petroleum industry, agriculture, and food services. Today, WLAN technology is becoming fairly mature, and it is becoming more widely recognized as a general-purpose connectivity alternative for a broad range of customers.

Still, the WLAN market remains small because the technology is new, so components are expensive and the data rates are low. For example, it costs less than $100 to buy a network card to connect a PC to a wired Ethernet LAN with a 10 Mbps data rate, but the card to interface the same PC to wireless radio LAN costs $500 and the wireless hub (access points) connecting the portable units to the wired network costs as much as $3000 for a data rate of 1–2 Mbps. However, research groups are working hard to shrink radios into a chip that can be mass produced cheaply. If they succeed, the demand for radio LANs may follow the same trend as cellular phones in recent years.

6.2 Wireless ATM

ATM, asynchronous transfer mode, technology is the result of efforts to devise a transmission and networking technology to provide high-speed broadband integrated services: a single infrastructure for data, voice, and video. Until recently, the integration of wireless access and mobility with ATM had received little attention.

Wireless ATM and other high-speed wireless networking technologies are motivated by the increasing demand for portable computing/telecommunications devices. In the early 1990s, researchers began to extend ATM capabilities into the wireless arena. Although wired ATM is becoming the technology of choice for broadband service integration, wireless ATM is an emerging technology with many issues still to be resolved.

The concept of wireless ATM (WATM) was first proposed in 1992. It is now regarded as the potential framework for next-generation wireless broadband communications that will support integrated, quality-of-service (QoS) multimedia services. Much research and development has been done, and many experimental wireless ATM network prototypes have been developed. WATM technology is now moving from the research stage to standardization and early commercialization.

This section briefly provides a background on ATM,[6–8] which is useful for understanding WATM.

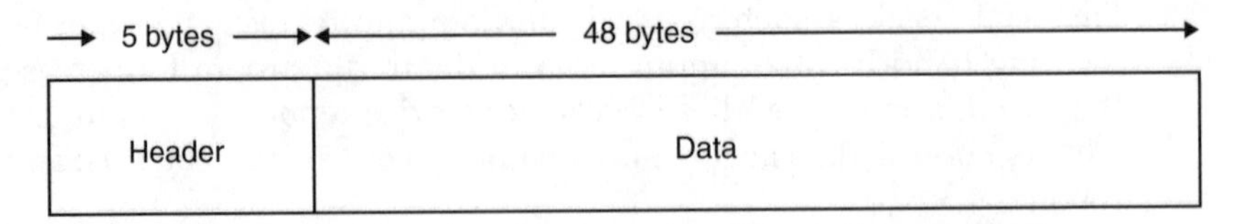

Figure 6.5 ATM cell format.

6.2.1 Overview of ATM

ATM is the switching and multiplexing ITU-T standard for broadband integrated services digital networks (BISDN). It has been advocated as the appropriate technology for wide area interconnection of heterogeneous wide area networks. It is a packet switching and multiplexing technique designed to handle voice, data, and video in a single physical channel. ATM is considered to be well suited as the transport mechanism for high-speed networks because of its support for bandwidth-intensive applications, its ability to carry different media types, and its ability to guarantee quality of service.

In an ATM network, the data is segmented into small fixed-length cells. The small fixed length makes it easier to bound the variation in delay across the system and also makes ATM networks easier to build than traditional packet-switching schemes. As shown in Figure 6.5, each cell is 53 bytes; 5 bytes serve as the header which comprises identification, control priority and routing information, and the remaining 48 bytes contain the actual data. (Each cell header, called a *label*, contains its destination address.) ATM cells are routed based on header information that identifies the cell as belonging to a specific ATM virtual connection. ATM supports bidirectional transfer of cells while preserving the order of transmission.

ATM networks are connection-oriented packet-switching networks designed for fiber optical links which are characterized by a large bandwidth and a very reliable almost error-free transmission. The ATM protocol requires that the QoS parameters be specified in the connection set-up phase. These parameters are:

- Peak-to-peak cell delay variation (CDV)
- Maximum cell transfer delay (CTD)
- Cell loss ratio (CLR)
- Cell error ratio (CER)

As a connection-oriented technique, a virtual circuit must be set up and established across the ATM network before any data can be transmitted from source to destination. An ATM connection is identified by two labels to the *virtual channel identifier* (VCI) and *virtual path identifier* (VPI). The relationships between a physical transmission path, VPIs, and VCIs are shown in Figure 6.6. The physical path can be a coaxial cable or fiber-based SONET. A physical path contains a certain number of virtual paths. Each virtual path has many virtual channels.

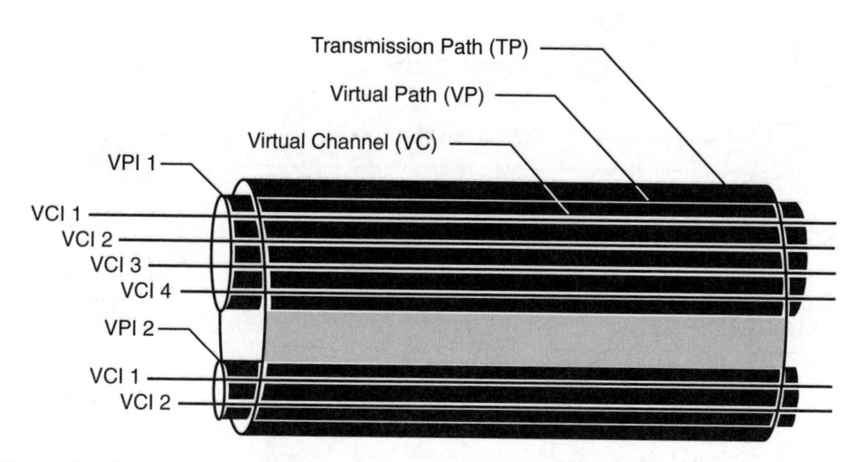

Figure 6.6 Relationship between transmission path (TP), virtual paths (VPs), and virtual channels (VCs).

An ATM network consists of ATM switches interconnected by point-to-point ATM links. The ATM switch constitutes an important component of the ATM network. A switch can handle several hundred thousand cells per second at each switch port. In its simplest form, a switch has a number of links to receive and transmit cells. The primary function of an ATM switch is to route cells from an input port to an appropriate output port at an extremely high bit rate. By interconnecting ATM switches, an ATM network can be developed to span a building, campus, city, nation, or the globe.

The ATM layer is implemented in both ATM endpoint devices and switches because the real task of transferring information takes place there. ATM switches support two kinds of interfaces: user-network interface (UNI) and network-network interface (NNI). The UNI describes the link between an ATM endpoint and an ATM switch. The NNI is a link that exists between two ATM switches or between two networks.

There are two types of ATM connections:

- Permanent virtual connections (PVC): Connection established by some external mechanism, in which a set of switches between an ATM source and destination ATM systems are programmed with the appropriate VPI and VCI values. PVCs always require some manual configuration.
- Switched virtual connections (SVC): A connection that is set up automatically through the signaling protocol. All higher layer protocols operating over ATM use primarily SVCs.

For each connection, there are four service classes identified as:

- Constant bit rate (CBR): Cells are generated by the source at a constant rate, e.g., pulse code modulated (PCM) speech.
- Variable bit rate (VBR): Cells are generated at a non-constant rate, e.g., video.

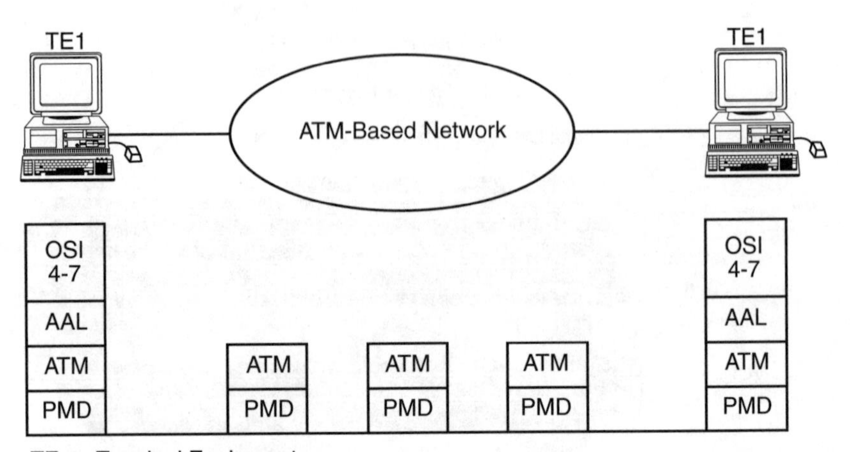

Figure 6.7 A typical ATM network.

- Available bit rate (AVR): Connection-oriented service utilizing the bandwidth remaining after CBR and VBR allocations.
- Unspecified bit rate (UBR): Packet-oriented, connectionless data without delay.

ATM networks are designed to support a wide range of applications with various service requirements and traffic requirements. ATM networks were initially envisioned as a transport technology for delivering integrated services on the public WANs in the form of BISDN. A typical ATM network is shown in Figure 6.7, where the terminal equipment (TEs) might be LANs, high-speed host computers, and CAD/CAM workstations. An ATM network allows the flexible allocation of bandwidth for packet, circuit-switched, or dedicated services.

6.2.2 Wireless ATM (WATM) architecture

The idea of WATM networks raises a number of challenges. First, integrated multimedia services demand high data rates (from 2 Mbps up), which are common in wired networks but still constitute a challenge for a wireless link. Second, ATM was designed with the assumption that the physical medium has a very low bit error rate (e.g., 10^{-9}), whereas wireless communications suffer from very high bit error rates (e.g., 10^{-3}) because of time-variant multipath propagation, blocking, and interference. Additional protocol layers and special coding techniques are required to share the radio channel bandwidth and minimize the bit error rate. Third, the incorporation of mobility in the connection-oriented BISDN requires the dynamic reestablishment of the ATM virtual circuits (VCs) with the short time interval the mobile terminal handover from one macrocell to another. In addition, the reestablishment of the VCs must ensure in-sequence and loss-free delivery of the ATMs in order to guarantee the QoS requirement on the connections. Moreover, the wireless channel is an expensive resource in terms of bandwidth, whereas ATM was designed for bandwidth-rich environment. ATM effectively trades off bandwidth for simplicity in switching. Every ATM cell carries a header with an overhead of about 10%. Even this much overhead is considered too high for a shared radio channel. If these challenges can be resolved, there will be significant advantages of wireless access.

The advantage of WATM networks is definitely mobility, resulting from the tetherless feature which cannot be supported by wired or fixed ATM networks. To enable nomadic access to ATM networks, however, additional functions are required in ATM networks for mobility support.[9]

WATM networks are basically the wireless extension of fixed ATM network. The 53-byte ATM cell is too big for WATM network; therefore, WATM networks can use 16- or 24-byte payload. Thus, in a WATM network, information is transmitted in the form of a large number of small transmission cells called *pico cells*. Each pico cell is served by a base station, while all the base stations in the network are connected via the wired ATM network. The ATM header is compressed or expanded to a standard ATM cell at the base station. Base stations are simple cell relays that translate the header formats from the WATM network to the wired ATM network. ATM cells are transmitted via radio frames between a central station (B-CS) and user radio modules (B-RM) as shown in Figure 6.8. All base stations operate on the same frequency so there is no hard boundary between pico cells.

Reducing the size of the pico cells helps in mitigating some of the major problems related to wireless LANs. The main difficulties encountered are the delay spread due to multipath effects and the lack of a line-of-sight path, which results in high attenuation. Also, small cells have some drawbacks compared to large cells.

From Figure 6.8, we notice that a wireless ATM typically consists of three major components: (1) ATM switches with standard UNI/NNI capabilities, (2) ATM base stations, and (3) wireless ATM terminal with a radio network interface card (NIC). There are two new hardware components: ATM base station and WATM NIC. The new software components are the mobile ATM protocol extension and WATM UNI driver.

In conventional mobile networks, transmission cells are "colored" using frequency-division multiplexing or code-division multiplexing to prevent interference between cells. Coloring is considered a waste of bandwidth because in order for it to be successful, before reusing a color an area of bandwidth must be idle. These inactive areas are wasted since they cannot be used for transmission.

WATM architecture is based on integration of radio access and mobility features. The idea is to integrate fully new wireless PHY, MAC, data link control (DLC), wireless control, and mobility signaling functions into the ATM protocol stack.

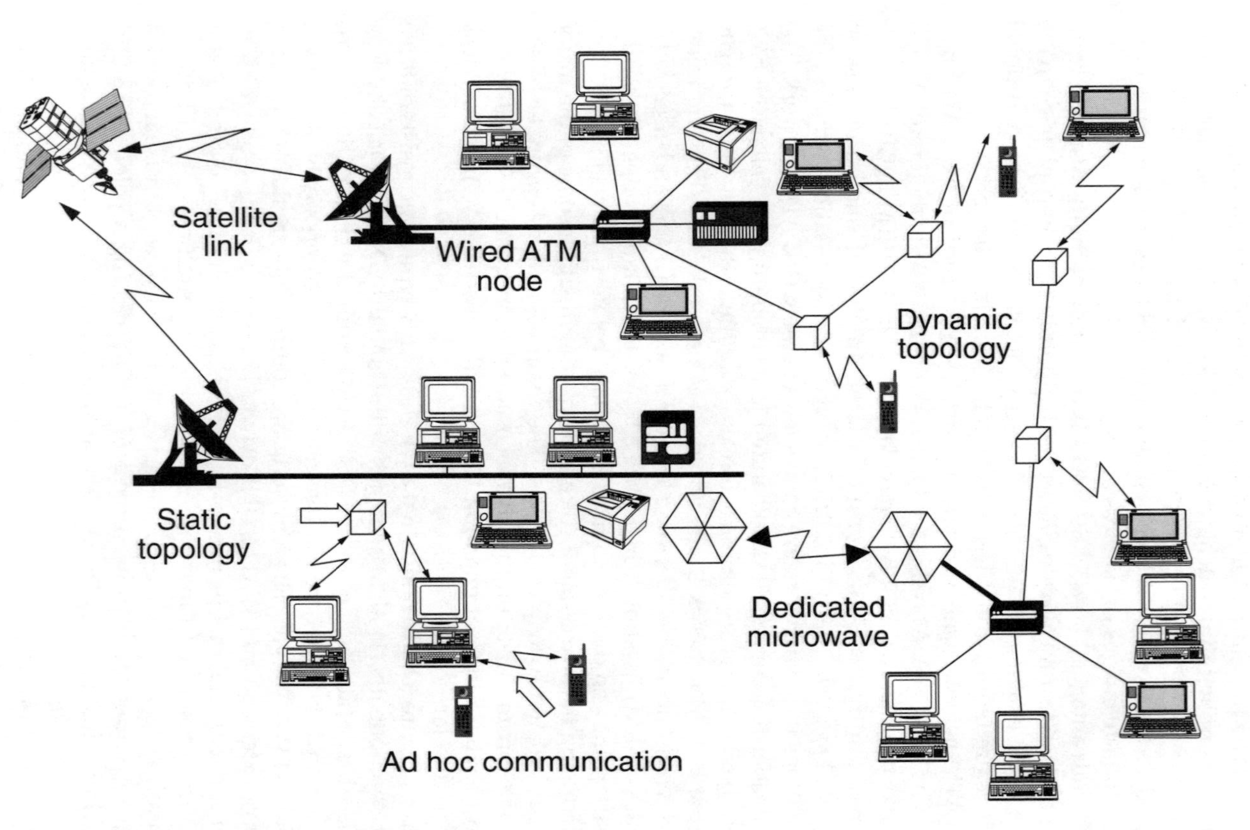

Figure 6.8 A typical wireless ATM network.

The requirements for the physical (PHY) layer include:

- *Frequency band:* It is expedient to use appropriate available band which is 5 GHz band for unlicensed in-building wireless ATM access.
- *Data rates:* High data rates are less essential for mobile devices with limited power and processing budgets. The ATM Forum has approved 25 Mbps as a standard PHY data rate for low cost ATM solutions.
- *Modulation efficiency:* Radio bandwidth is a rare resource. It is therefore important, especially at 5 GHz, to utilize an efficient modulation technique.

As for the medium access control (MAC), a WATM needs to have some method of accessing the shared channel. The requirements in a WATM system are very stringent as the MAC protocol needs to provide guarantees for both asynchronous and isochronous traffic types in the face of an unreliable channel. The MAC protocol must also facilitate efficient hand-off between base stations for mobile units and allocate the available bandwidth in a flexible and dynamic manner. Since one of the major advantages of ATM is its ability to give QoS guarantees to connections, the MAC protocol of a WATM network should support all the service categories including constant bit rate (CBR), available bit rate (ABR), and unspecified bit rate (UBR). Also, since many WATM devices are portable and battery powered with limited power budgets, it is important to consider power-saving techniques in all radio access protocol layers.

For the DLC, the reliability of the radio channel needs to be increased using link-level error-control procedures. Techniques to recover from bit errors include the use of forward error control (FEC) and cyclic redundancy check (CRC) in combination with an automatic repeat request (ARQ). As mentioned earlier, the ATM protocol was designed with the assumption of highly reliable links and point-to-point transmission, which contrasts with the characteristics of the wireless link. To overcome these incompatibilities, the DLC layer is introduced between the ATM layer and the physical layer, as shown in Figure 6.9. WATM will need a custom data link layer protocol due to the high error rate and different packet size for WATM. As mentioned

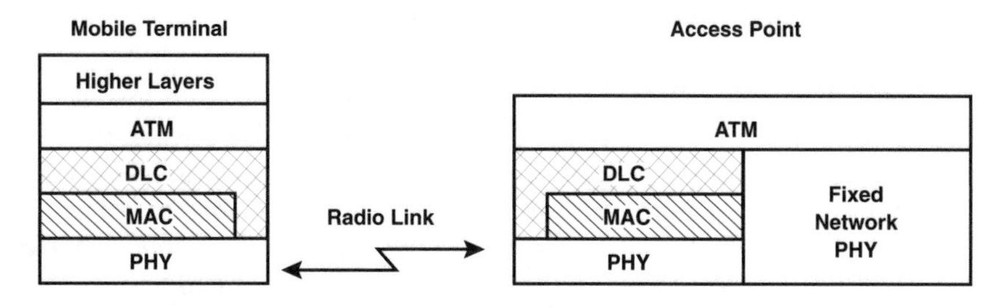

Figure 6.9 Wireless ATM protocol architecture.

earlier, WATM can use a 16- or 24-byte payload since 53 bytes might be too long. The data link protocol can contain the service type definition, error control, segmentation and reassembly, and hand-off support. The service type will indicate whether a cell is of type control, CBR, VBR, etc. Error control is needed because of interference and the poor physical-level characteristics of the physical medium. Segmentation and reassembly is required because WATM can use 16- or 24-byte cells. Hand-off, occurring when the mobile unit moves from one area to another, should be transparent.

Because eavesdropping and spoofing are generally harder to detect in wireless systems than in wired, security must be carefully considered. Encryption must be included at the physical layer, which will, of course, increase hardware cost. Also, a set of requirements must be defined for the hand-off process. The hand-off protocol enables a wireless terminal or wireless network to move seamlessly between access points (AP) while retaining the negotiated QoS of its connections. This can be achieved by the efficient rerouting and switching of the wireless terminal's VCs to the appropriate AP in the wired ATM network.

Potential applications of wireless ATM networks include:

- Private ATM LANs
- Portable enhanced television services
- Wireless Internet access and mobile network computers
- Cordless telephony
- Tactical networks

Wireless mobile ATM (WMATM) is evolving from WATM. It is extending broadband multimedia services to mobile wireless terminals. It provides innovative solutions to constructing a generic broadband wireless core to ensure wireless quality of service and mobility management.[10]

WATM is not as mature as wired LAN. No standards have been defined by either ITU-T or the ATM Forum. However, the ATM Forum's WATM Working Group (started in June 1996) is developing specifications that will facilitate deployment of WATM. It is clear that commercially available WATM systems are in the not-too-distant future. More information about WATM can be found in References 11 through 14.

6.3 *Wireless local loop*

The wireless local loop (WLL), also known as fixed-wireless services, is now being used in place of conventional wire-line connections between the local exchange and customer premises. Conventionally, the local loop or access network consists of a pair of copper wires, buried or suspended in air, connecting subscribers at home or office to the nearest telephone exchange. It is typically 6 to 8 km long, and the copper gauge used is 0.5 to 0.6 mm. Local loop is designed to carry 4 kHz voice and is difficult to maintain, with

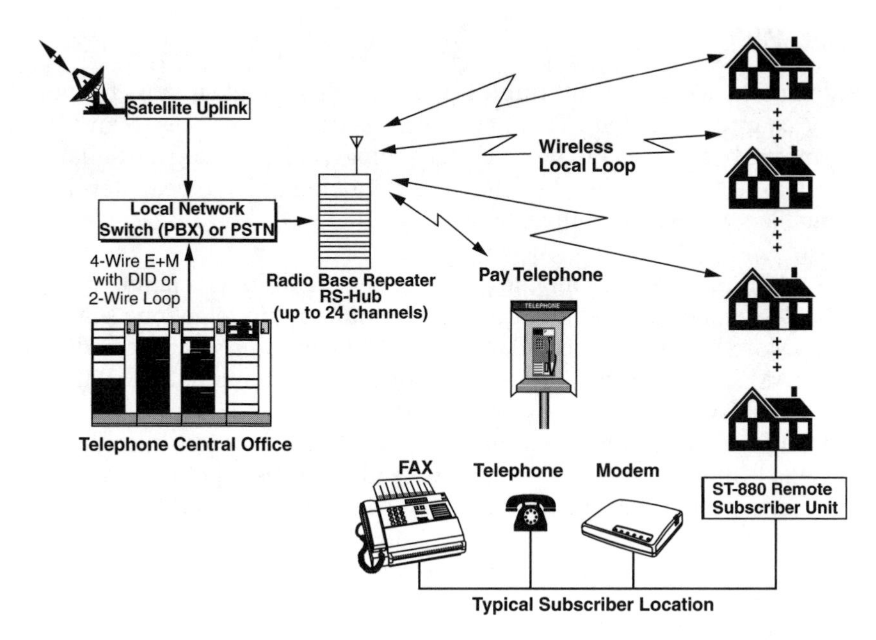

Figure 6.10 A typical wireless local loop system.

85% of all faults found in the local loop. It is expensive and time-consuming to deploy in view of the rising cost of copper and digging.[2]

WLL, also known as radio in the loop (RITL) or fixed-radio access (FRA), eliminates the physical connection between telephone exchange and subscribers. WLL connects subscribers to their local exchange using radio signals in place of conventional copper cabling. This includes cordless access systems, proprietary fixed-radio access, and fixed-cellular systems. A typical WLL system is shown in Figure 6.10.

The principle of the WLL is simple. A radio base station plays the role of transmitter-receiver and enables connection with houses, equipped with a simple antenna, within a radius of 3 km. The base portion of the system is the radio base station hub which is located near the telephone central office. The wireless portion of the system is the distance between the hub and the individual subscribers, each of whom has a full duplex transceiver. A terminal and some antennas are thus enough to constitute the architecture of these networks forming a wireless local loop.

While the local loop or access network is fully in place in some regions of the world, it remains to be deployed in remote or rural areas where half of the world's population lacks plain old telephone service (POTS). Developing nations such as China, India, Brazil, Indonesia, and the Philippines look to WLL technology for an efficient way to deploy POTS without the expense of burying tons of copper wire. In developed nations, WLL will enable new operators to bypass existing wire-line networks to deliver POTS and data access.

6.3.1 WLL services

By definition, WLL is a full duplex, voice grade service suitable for voice, fax, or computer modem. There are varieties of WLL services. Many WLL systems are based on cellular or PCS technology, either analog or digital, but WLL is fixed, not mobile. The following technologies have been developed for mobile wireless communications:

- AMPS: Advanced mobile phone service (AMPS) is a family of wireless standards intended to be just another radio-telephone standard. More than half the cellular phones in the world operate according to AMPS standards. AMPS has grown to accommodate TDMA and CDMA.
- TDMA: Time division multiple access (TDMA) divides a single channel into a number of time slots, with each user getting one out of every few slots. For example, TDMA, as defined in IS-54 and IS-136 standards, divides a 30-kHz cellular channel into three time slots, which supports three users in strict alternation. TDMA is the most widely used multiple access technique for WLL.
- CDMA: Code division multiple access (CDMA) is a spread spectrum technology, which means that it spreads the information contained in a particular signal over a bandwidth much greater than the original signal. A call starts with a rate of 9.6 kbps and spreads to about 1.24 Mbps. *Spreading* means that digital codes are applied to the data bits associated with users in a cell. When the signal is received, the codes are removed and the call is returned to a rate of 9.6 kbps. Unlike an analog conversation, CDMA calls are secure from the casual eavesdropper.
- PCS: Personal communications systems (PCS) is a wireless system operating in the 1.8 GHz frequency band. WLL applications typically take advantage of other existing wireless technology, such as cellular and PCS system architecture, to reduce cost. To date, no international frequency is allocated for WLL systems. In many cases, mobile cellular or PCS systems compete with fixed WLL systems for shared radio spectrum resources.
- GSM: The global system for mobile telecommunications (GSM) systems are also offered. Initially designed for 900 MHz operation, the GSM systems are now available in 1800 MHz and 1900 MHz. GSM is the biggest challenger to AMPS technology and is the most dominant system used in the world. However, since GSM's architecture was designed to handle international roaming, it carries large overhead that makes it costly for WLL applications. Despite this limitation, GSM-based WLL products may be developed in the future.

More than two dozen WLL technologies are on the market today, many offered by telecom companies. They can be divided into two categories:

Table 6.1 Two Types of WLL Applications

High Tele-Density	Low Tele-Density
Multiple low-power cell sites	Single high-power base station
Usually cellular or PCS technology	Narrowband FM
Usually 800/900 MHz or higher	Usually below 800 MHz
Low power remote equipment	Mid- to high-power remote equipment
Broadband FM — 200 kHz channels	Narrowband — 25 kHz channels
Limited range	Wide range

- Wireless systems that use radio links for the fixed telephone services: CT2 (cordless telephone, generation 2), DECT (digital enhanced cordless telecommunications), and PHS (personal handyphone system) — CT2 and DECT are European standards, and PHS is a Japanese standard.
- Systems that use mobile cellular technology, such as DCS (digital communications services), derived from GSM and CDMA.

6.3.2 WLL applications

Tele-density refers to the concentration of subscribers within a WLL service area. A large urban area with many subscribers is considered to have a high tele-density, whereas a rural area with few subscribers would have low tele-density. Table 6.1 compares high and low tele-density applications. The high-density applications tend to favor cell-based systems using cellular or PCS technology. The cost of the infrastructure tends to be relatively high due to the need to construct and maintain multiple cell sites. On the other hand, low tele-density applications are characterized by a single, high-power hub located in the center of the covering area. Due to the low cost of the hub equipment, trunked mobile radio technology can be very attractive for lower tele-density applications.

As mentioned earlier, the great potential of WLL applications can be found in developing nations where wire-line infrastructure is either nonexistent or inadequate to meet the growing demands. These countries are experiencing strong economic growth which in turn creates a strong demand for communications services.

WLL technology is said to offer a number of key advantages over traditional connections based on copper cables. These include:

- *Faster deployment:* It takes weeks or months to deploy WLL systems compared with months or years needed to deploy underground or above-ground copper wire.
- *Lower construction costs:* It takes less heavy construction to deploy WLL technology compared with the labor-intensive installation for laying of copper lines.

- *Lower operating/maintenance costs:* Wireless equipment is less failure-prone than copper and can be less vulnerable to theft or damage.
- *Lower network extension costs:* Once the WLL infrastructure is in place, each incremental subscriber can be installed at very little cost.

Other advantages include faster realization of revenues, reduced time to payback of the deployment, fast network expansion of subscribers, lower network maintenance, and improved reliability.

Despite these advantages, the number of telephone lines currently supported by WLL technology remains relatively small. The key factors preventing the rapid deployment of WLL include lack of world-wide frequency allocation and technical standards. The use of a technology shared with mobile services becomes a disadvantage in the deployment of a WLL system despite the cost savings. Also, WLL equipment manufacturers, lacking an international frequency allocation, cannot generate the volume needed to realize reduced costs. However, planned and potential deployments offer the prospect of significantly larger systems. More about WLL can be found in Reference 15.

6.4 Wireless PBXs

The wireless private branch exchange (WPBX) technology is the core of the wireless office. Workers are being forced by job demands to be mobile but not far from their telephones. WPBX technology is designed to meet both demands. WPBX replaces the traditional desktop phone with mobile handsets, allowing employees to roam with their desktop and still have full PBX functionality available on their mobile handsets. This situation results in increased productivity — no missed calls and fewer phone calls requiring to be returned.

The development of the wireless PBX technology is driven by a number of needs, including:[2]

- *Time sensitivity:* immediate access to vital information
- *Individualization:* customized communication systems so users have ready access to communication capabilities
- *Personal mobility and portability:* ability to fulfill time-sensitive, work-related requirements while untethered from a fixed network

WPBX is particularly useful in multibuilding sites in a campus environment, e.g., manufacturing, business, government, education, and health care.

Wireless PBX has its roots in standards developed in Europe and Japan. The following three standards are bidding for attention in the wireless PBX:

- *The cordless telecommunication (CT) series of standards:* These are low-cost, public cordless telephone systems that have been popular in Europe. CT has generations 0, 1, 2, and 3 corresponding to CT0, CT1,

CT2, and CT3, respectively. CT0/CT1 are for analog cordless telephone. CT2 allows bandwidth splitting into radio channels. CT3 was designed for high-density office environments, using technology to boost capacity in crowded areas. To ensure compatibility with other communication devices, many wireless systems adhere to CT2.

- *The digital enhanced cordless telecommunications (DECT):* DECT is a Pan-European standard for digital cordless telephony which operates from 1880 MHz to 1900 MHz.
- *The personal handy-phone system (PHS):* PHS is a Japanese standard that uses TDMA/TDD system and occupies the 1895-1918.1 MHz band. (The spectrum bands for DECT and PHS overlap slightly.)

For example, a WPBX may use PHS technology. Such a WPBX provides the same capabilities as a public PHS system, including automatic location registration, hand-off, and authentication. It also allows the use of many functions inherent in a PBX. System capacity depends on the capacity of the PBX that is used. One initial advantage of PHS-based WPBX systems over DECT, claimed by Japanese vendors, is the ability to extend the reach of the PBX beyond the office. PHS lost its edge in providing this ability because of the lack of public PHS mobile networks outside of Japan. DECT-based WPBX systems had the same basic problem.

The topology of a typical wireless PBX is shown in Figure 6.11. A wireless PBX can consist of the following components:

- *Modular controller:* an adjunct switch connecting the twisted-pair wiring or optical fiber to PBX. It contains the CPU and control logic and manages the calls sent and received between the base stations.
- *Base stations:* strategically placed units that relay the signals via antennas to individual phones. About the size of smoke detectors, they are mounted on the ceiling and are connected to the wireless PBX via twisted-pair wiring. Several base stations are connected to a PBX, and the PBX is connected to the public switched telephone network (PSTN).
- *Telephone handsets:* portable, pocket-sized, digital telephone. Each handset must be registered with the adjunct switch to limit only authorized users to access the communications system.
- *Distribution hub:* used in a large installation, a distribution hub acts as a "traffic cop" between multiple base stations within a cell. It allows high-traffic locations to be divided into smaller cells. It is connected to the adjunct unit via twisted-pair wiring or optical fiber.

Major PBX vendors (such as Ericsson, Lucent, Nortel, and Siemens) offer wireless extensions optimized for the campus environment. The extensions are wireless switches that are connected to an existing PBX and Centrex. This arrangement provides wireless telephone, paging, and e-mailing services to nomadic employees within the workplace. By being able to communicate

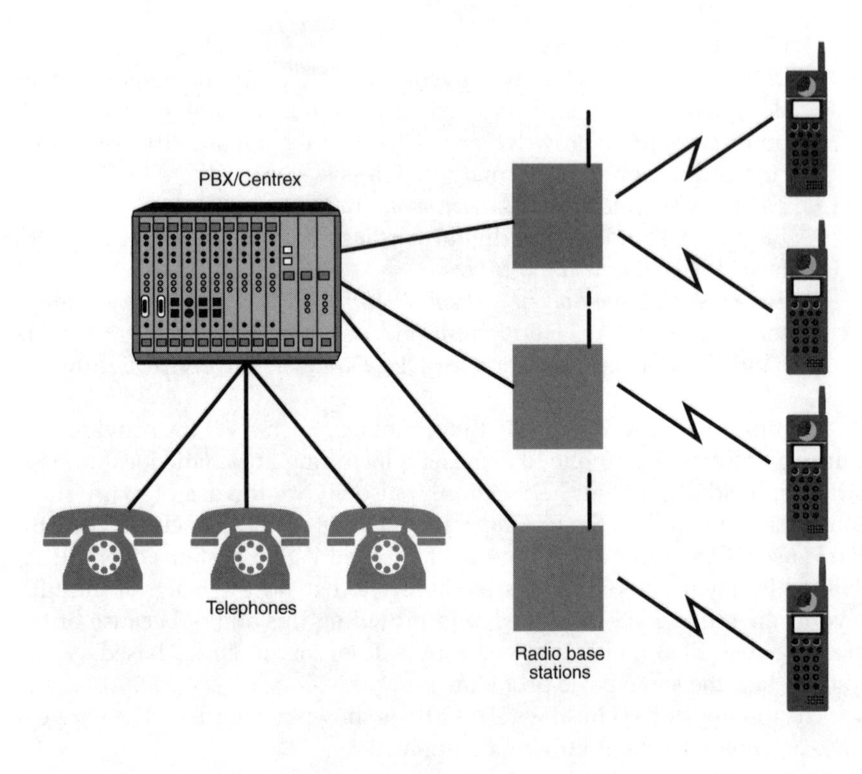

Figure 6.11 The topology of a typical wireless PBX.

on-the-go, employees and managers can make better use of their time and provide customers better services. WPBX is used indoors and gets better reception and longer battery life.

Wireless PBXs operate in different frequency bands, including the unlicensed PCS band of 1910 to 1930 MHz. Since a WPBX operates over such bands in a very small geographical area, it has little or no interference with other wireless services in the area. Therefore, there is no need for FCC licensing.

While the wireless PBXs have been in the marketplace for over a decade, they have yet to make a dent in the market. The WPBX industry is in a "Catch -22" — costs will not come down until volume increases, but current prices limit demand. Consequently, vendors are hoping that new capabilities will help stimulate the market.

6.5 Wireless PAN

Wearable computer devices, such as cellular phones, personal digital assistants (PDAs), pagers, personal stereos, pocket video games, and notebook computers are getting smaller and less expensive. There is an ever-increasing

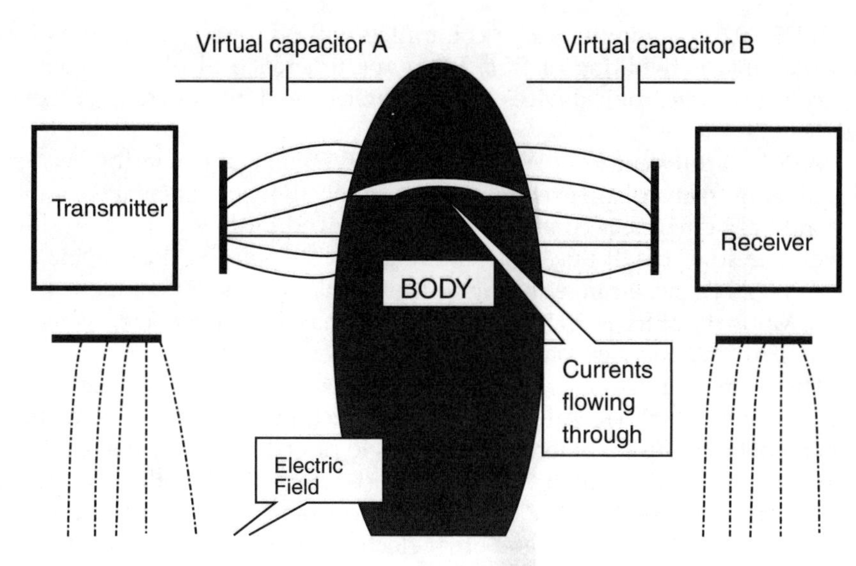

Figure 6.12 A wireless personal area network.

use of these personal devices and the number of these devices is increasing as well. However, there is currently no means for these devices to share or exchange data among themselves. To facilitate communications among these personal devices, a new wireless communications technology, *wireless personal area network*, is emerging.[16]

Wireless personal area network (WPAN) enables communication or information exchange between computing devices that can be portable or mobile and can be worn or carried by individuals. WPAN operates in the license-free radio frequency band of 2.4 GHz (2400 to 2483.5 MHz), with ranges up to 10 m and data rates up to 1 Mbps. It ties together closely related objects, a function that is fundamentally different from the objective of a WLAN. Thus, WPAN must coexist with other radio technologies, such as WLAN, that operate in the same frequency band.

WPAN can be regarded as a "body network." It uses a near-field electric field to send data across various devices using the electrical conductivity of the human body as a medium. As shown in Figure 6.12, WPAN enables communication between two devices using the human body as a medium by using a transmitter and receiver, both battery powered. The transmitter, which need not be in direct contact with the skin, is capacitively coupled through the body to the receiver. The system electrostatically induces picoamp currents (too little to be felt by the body or to have any biological effects) into the body, which is used to conduct the modulated currents. A typical prototype is slightly larger and thicker than a credit card and consumes 1.5 mW to power its coupling electrodes, which are not in direct contact with the skin but placed in shoes, wallets, pockets, etc.

WPAN was developed at the M.I.T. Media Laboratory and later supported by IBM Almaden Research Laboratory. It is being standardized by the IEEE 802.15 committee. The committee will address requirements for personal area networking of PCs, PDAs, peripherals, cell phones, pagers, and consumer electronic devices to communicate and interoperate with one another.

WPAN is different from WLAN. First, WPANs target primarily the vast consumer market and are used for ease-of-connectivity of personal wearable or hand-held devices. Second, WPAN is optimized for low complexity, low power, and low cost. It does not require an access point as does a WLAN. Third, WPAN's close range throughput of 1 Mbps does not compare with the 11 Mbps the IEEE 802.11 WLAN offers. It has small coverage, typically about 10 m, and connects only a limited number of devices.

The terms WPAN and Bluetooth have been used interchangeably in many technical articles. Although the two technologies are similar, they should not be confused. Bluetooth is a far-field radio technology that is being promoted by the Bluetooth special interest group (SIP) with more than 75 members. Bluetooth technology is for short-range, low-cost radio links between PCs, mobile phones, and other electronic devices. It provides a 10-m personal bubble that supports simultaneous transmission of both data and voice for multiple devices.

Some interesting applications of PAN have been proposed, including:

- Transfer of information between all the electronic devices a gadget-hungry person might carry. Such devices include pagers, cellular phones, PDAs, identification badges, and smart cards.
- Exchange of business cards by just shaking hands. The electronic card is transferred automatically from one card device via the body to the other person's card device.
- Office and domestic automation. One is able to control lighting, heating, and even locks with a gentle touch. These devices and appliances are programmed to communicate with each other with little or no human intervention.
- Telemedicine with intelligent sensors monitoring specific physiological signals, such as EEG, ECG, and GSR, and performing data acquisition, as shown in Figure 6.13. Such a collection of wearable medical sensors can communicate using PAN.[17]

Other potential applications include sensor and automation needs, interactive toys, and location tracking for smart tags and badges.

A major problem with PAN is security. Since transferring data across devices has become simple, much work on encryption needs to be done.

PAN is in the prototyping and standardization stage today, and most of the work has been done in the lower layers. More work needs to be done at the higher layers to resolve issues such as security, authentication, and reliability. Like any other technology, the success of PAN does not depend on what it promises to achieve but on what it is actually able to deliver.

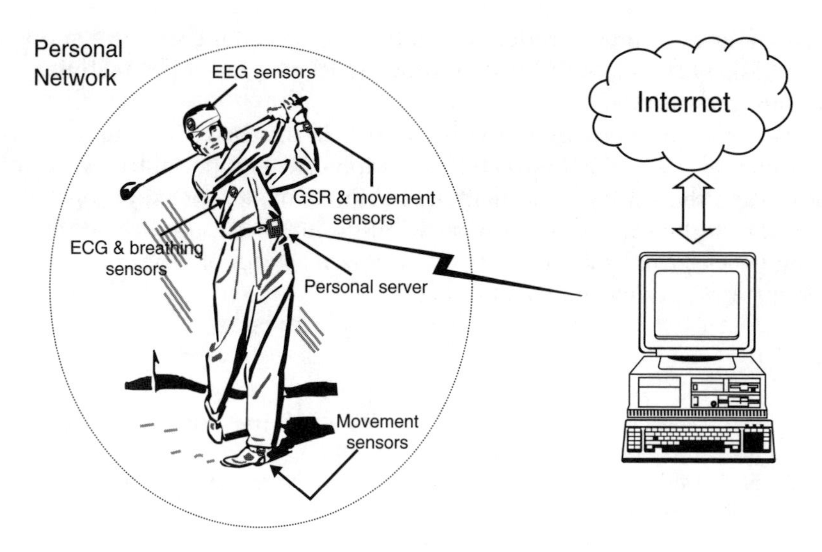

Figure 6.13 A typical wireless PAN for medical application.

Summary

Wireless communication networks include everything from cellular, personal communications system (PCS), and group system for mobile communications (GSM) networks to wireless LANs (WLAN), wireless ATM (WATM), and satellite-based networks. Although wireless data communication is one of the most promising communications technologies, the industry is still in its infancy.

WLAN allows laptop PC and LAN users to link through radio waves or infrared links, eliminating the need for restrictive cables and opening a wider range of possible applications.

More recently, much attention has been focused on the extension of ATM technology to wireless communications. In view of the emerging multimedia applications and the demand for mobility, it is apparent that WATM networks will become important in the future. Target data rates of up to 155 Mbps are envisioned for WATM. The standardization process of WATM is in progress. The rapid growth of wireless communications coupled with the remarkable development in ATM networking signals the beginning of a new era in telecommunications.

WLL technology is becoming a cost-effective solution for local telephone network access. It is only one of the many competing technologies that can be used to provide local-loop access. It replaces the last few hundred meters of the access network cable that runs to the end user's home with a radio system. The potential market for WLL is enormous, especially in developing nations.

The idea of wireless PBX is to facilitate communication within the office environment and to enable workers to be as productive with a wireless handset, or even more productive, as they would be if they were sitting at their desks. A wireless PBX can improve efficiency, boost productivity, and increase responsiveness.

Personal area network (PAN) is a near-field intrabody communication. It is a wireless connection between PCs, peripherals, and portables that will let the devices share information without having to make a physical connection.

More technical information on wireless communication networks and related issues can be obtained from the special series of the *IEEE Journal of Selected Areas in Communications*.[18–27]

References

1. P. T. Davis and C. R. McGuffin, *Wireless Local Area Networks*, McGraw-Hill, New York, 1995, 41–117.
2. N. J. Muller, *Mobile Telecommunications Factbook*, McGraw-Hill, New York, 1998, 219–270.
3. V. K. Garg, K. Smolik, and J. E. Wilkes, *Applications of CDMA in Wireless/Personal Communications*, Prentice-Hall, Upper Saddle River, NJ, 1997, 233–272.
4. F. J. Ricci, *Personal Communications Systems Applications*, Prentice-Hall, Upper Saddle River, NJ, 1997, 109–118.
5. G. Anastasi et al., MAC protocols for wideband wireless local access: Evolution toward wireless ATM, *IEEE Personal Commun.*, Oct. 1998, 53–64.
6. M. N. O. Sadiku, *Metropolitan and Wide Area Networks*, Prentice-Hall, Upper Saddle River, NJ, in press.
7. M. Barton et al., Wireless ATM: Internetworking Aspects, in J. B. Gibson (Ed.), *The Mobile Communications Handbook*, 2nd ed., CRC Press, Boca Raton, FL, 1999, 34-1 to 34-21.
8. P. Pattulo and R. Steele, Wireless ATM, in R. Steele and L. Hanzo (Eds.), *Mobile Radio Communications*, 2nd ed., John Wiley & Sons, Chichester, 1999, 965–1043.
9. M. Umehira et al., ATM wireless access for mobile multimedia: Concept and architecture, *IEEE Personal Commun.*, Oct. 1996, 39–48.
10. *IEEE Comm. Mag.*, Sept. 1999, special issue on wireless mobile and ATM technologies for third-generation wireless communications.
11. *IEEE Comm. Mag.*, Nov. 1997, special issue on mobile and wireless ATM.
12. *IEEE J. Selected Areas Commun.*, vol. 15, no. 1, January 1997, special issue on Wireless ATM.
13. *IEEE Personal Comm.*, August 1996, special issue on Wireless ATM.
14. D. Raychaudhuri, Wireless ATM Networks: Technology status and future directions, *IEEE Proc.*, vol. 87, no. 10, Oct. 1999, 1790–1806.
15. W. Webb, *Introduction to Wireless Local Loop: Broadband and Narrowband Systems*, 2nd ed., Artech House, Boston, 2000.
16. T. M. Siep and I. C. Gifford, Wireless personal area network communications: an application overview, in M. Golio (Ed.), *The RF and Microwave Handbook*, CRC Press, Boca Raton, FL, 2001, 2-80 to 2-91.
17. E. Jovanov, Wireless personal area network in telemedical environment, *Proc. IEEE EMBS Int. Conf. Inform. Tech. Applications Biomedicine*, 2000, 22–27.
18. Wireless Communication Series, *IEEE J. Selected Areas Commun.*, vol. 17, no. 3, March 1999.

19. Wireless Communication Series, *IEEE J. Selected Areas Commun.*, vol. 17, no. 7, July 1999.
20. Broadband Wireless Techniques, *IEEE J. Selected Areas Commun.*, vol. 17, no. 10, Oct. 1999.
21. Wireless Communication Series, *IEEE J. Selected Areas Commun.*, vol. 17, no. 11, Nov. 1999.
22. Wireless Communication Series, *IEEE J. Selected Areas Commun.*, vol. 18, no. 3, March 2000.
23. Wireless Communication Series, *IEEE J. Selected Areas Commun.*, vol. 18, no. 7, July 2000.
24. Wireless Communication Series, *IEEE J. Selected Areas Commun.*, vol. 18, no. 11, Nov. 2000.
25. Wireless Communication Series, *IEEE J. Selected Areas Commun.*, vol. 19, no. 2, Feb. 2001.
26. Wireless Communication Series, *IEEE J. Selected Areas Commun.*, vol. 19, no. 6, June 2001.
27. Wireless Communication Series, *IEEE J. Selected Areas Commun.*, vol. 18, no. 7, July 2001.

Problems

6.1 Compare and contrast CSMA/CD and CSMA/CA.

6.2 Compare and contrast RF LAN and IR LAN.

6.3 Describe an ATM cell structure. What is the percentage of overhead?

6.4 At a user-network interface (UNI), how many virtual paths can exist in a physical path? How many virtual channels can exist in a virtual path?

6.5 What does a VPI identify?

6.6 What are the challenges facing wireless ATM networks?

6.7 Describe the requirements for the PHY, MAC, and DLC layers of a wireless ATM network.

6.8 Discuss some problems preventing the rapid deployment of WLL.

6.9 Describe a wireless PBX.

6.10 How is a WPAN different from a WLAN?

chapter seven

Cellular technologies

What you do speaks so loudly that I cannot hear what you say.

— Ralph Waldo Emerson

Mobile and wireless systems are not the same, although there is considerable overlap. Mobile networks provide support for routing (how to maintain communication with mobility) and location management (keeping track of the location) functions. Wireless networks provide wireless interfaces to users (both mobile and stationary). When mobility and wireless are combined, interesting issues arise that bear on performance, interoperability, and competing standards.

Perhaps no single development has done more for wireless technologies than has cellular communication. It is one of the fastest growing and most demanding telecommunication applications. Currently, there are more than 45 million cellular subscribers worldwide, and nearly half of them are in the U.S. It has been predicted that cellular will become the universal method of communication.

The cellular concept is not so much a new technology as it is a new idea of organizing old technology. It was developed in 1947 at Bell Laboratories; the first cellular system began operation in Japan in 1979, and the first cellular system in the U.S. began in October, 1983 in Chicago. The first generation of cellular systems was based on analog FM radio technology. Several analog cellular systems were in the U.S., Europe, and Japan without a worldwide standard. Limitations of analog FM radio technology led to the development of second-generation cellular systems which are based on digital radio technology. The second-generation digital cellular systems conform to at least three standards: GSM for Europe and international applications, AMPS for the U.S., and JDC for Japan. Third-generation cellular systems use TDMA, CDMA, CSMA, and FDMA. They are targeted to offer a wide variety of services, such as wireless extensions of ISDN.

This chapter examines different cellular technologies, including personal communication service (PCS) and cellular digital packet data (CDPD).

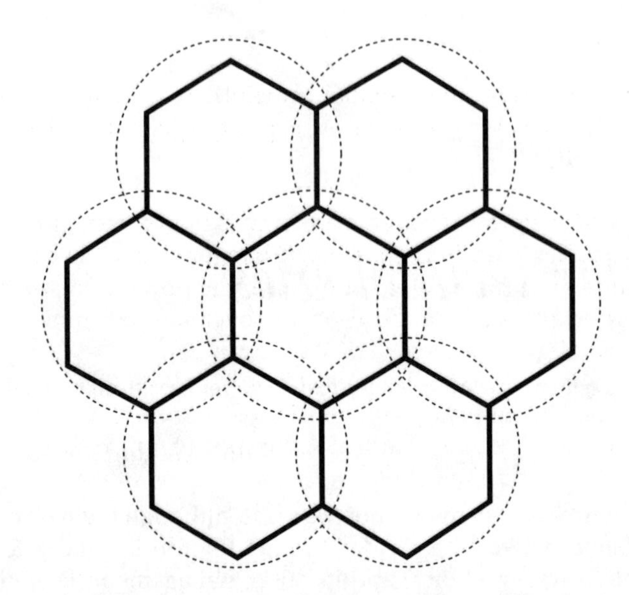

Figure 7.1 A typical wireless seven-cell pattern; cells overlap to provide greater coverage.

7.1 The cellular concept

The conventional approach to mobile radio involved setting up a high-power transmitter on top of the highest point in the coverage area. The mobile telephone had to have a line-of-sight to the base station for proper coverage. Line-of-sight transmission is limited to as much as 40 to 50 mi on the horizon. Also, if a mobile travels too far from its base station, the quality of the communications link becomes unacceptable. These and other limitations of conventional mobile telephone systems are overcome by cellular technology.

Areas of coverage are divided into small hexagonal radio coverage units known as *cells*. A cell is the basic geographic unit of a cellular system. A cellular communications system employs a large number of low-power wireless transmitters to create the cells. These cells overlap at the outer boundaries, as shown in Figure 7.1. Cells are base stations transmitting over small geographic areas that are represented as hexagons. Cell size varies depending on the landscape and tele-density. (The towers one sees on hilltops with triangular structures at the top are cellular telephone sites.) Each site typically covers an area of 15 mi across, depending on the local terrain. The cell sites are spaced over the area to provide a slightly overlapping blanket of coverage. Like the early mobile systems, the base station communicates with mobiles via a channel. The channel is made of two frequencies: the *forward link* for transmitting information to the base station and the *reverse link* to receive from the base station.

7.1.1 Fundamental features

Besides the idea of cells, the essential principles of cellular systems include cell splitting, frequency reuse, hand-off, capacity, spectral efficiency, mobility, and roaming.[1]

- *Cell splitting:* As a service area becomes full of users, the area is split into smaller ones. Consequently, urban regions with heavy traffic can be split into as many areas as necessary to provide acceptable service, while a large cell can be used to cover remote rural regions. Cell splitting increases the capacity of the system.
- *Frequency reuse:* This is the core concept that defines the cellular system. The cellular-telephone industry is faced with a dilemma: services are growing rapidly and users are demanding more sophisticated call-handling features, but the amount of the electromagnetic (EM) spectrum allocation for cellular service is fixed. This dilemma is overcome by the ability to reuse the same frequency (channel) many times. Several frequency reuse patterns are in use in the cellular industry, each with its advantages and disadvantages. A typical example is shown in Figure 7.2, where all available channels are divided into 21 frequency groups numbered 1 to 21. Each cell is assigned three frequency groups. For example, the same frequencies are reused in the cell designated as 1 and adjacent locations do not reuse the same frequencies. A cluster is a group of cells; frequency reuse does not apply to clusters.
- *Hand-off:* This is another fundamental feature of the cellular technology. When a call is in progress and the switch from one cell to another becomes necessary, a hand-off takes place. Hand-off is important, because adjacent cells do not use the same radio channels as a mobile user travels from one cell to another during a call, the call must be either dropped or transferred from one channel to another. Dropping the call is not acceptable. Hand-off was created to solve the problem. Handing off from cell to cell is the process of transferring the mobile unit that has a call on a voice channel to another voice channel, all done without interfering with the call. The need for hand-off is determined by the quality of the signal, whether it is weak or strong. A hand-off threshold is predefined. When the received signal level is weak and reaches the threshold, the system provides a stronger channel from an adjacent cell. This hand-off process continues as the mobile moves from one cell to another as long as the mobile is in the coverage area. A number of algorithms are used to generate and process a hand-off request and eventual hand-off order.
- *Mobility and roaming:* Mobility implies that a mobile user while in motion will be able to maintain the same call without service interruption. This is made possible by the built-in hand-off mechanism

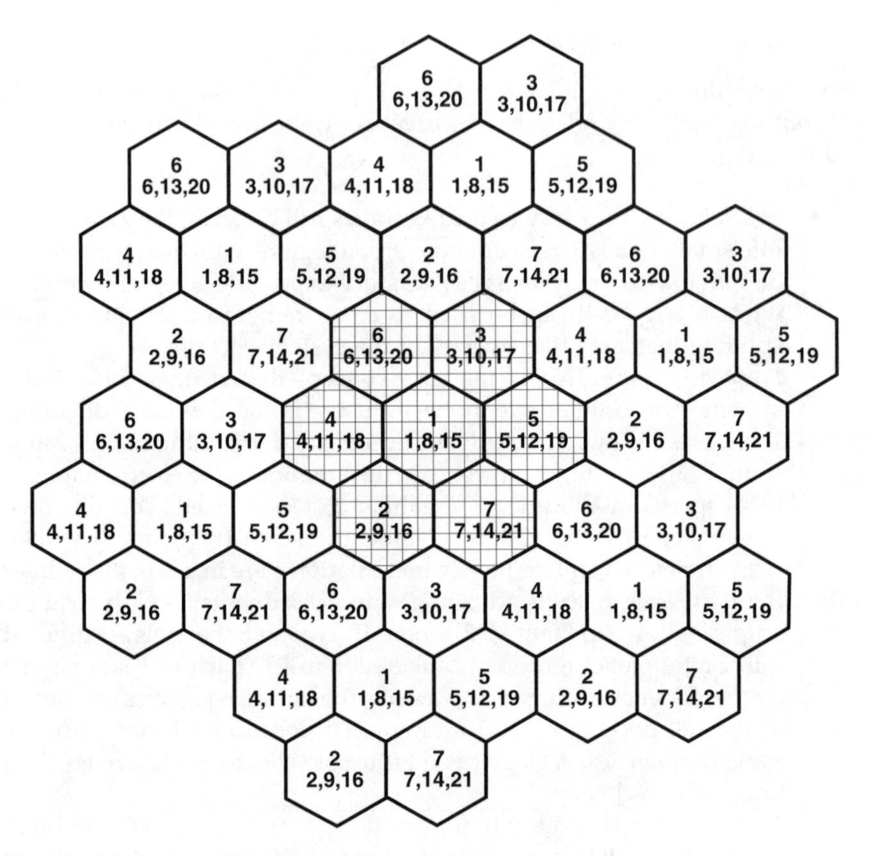

Figure 7.2 Frequency reuse in a seven-cell pattern cellular system.

that assigns a new frequency when the mobile moves to another cell. Because several cellular operators within the same region use different equipment, and a subscriber is registered with only one operator, some form of agreement is necessary to provide services to subscribers. Roaming is the process whereby a mobile moves out of its own territory and establishes a call from another territory.

- *Capacity:* This is the number of subscribers that can use the cellular system. For an FDMA system, the capacity is determined by the loading (number of calls and the average time per call) and system layout (size of cells and amount of frequency reuse utilized). Capacity expansion is required because cellular systems must serve more subscribers. It takes place through frequency reuse, cell splitting, planning, and redesigning of the system.
- *Spectral efficiency:* This is a performance measure of the efficient use of the frequency spectrum. It is the most desirable feature of a mobile communication system. It produces a measure of how efficiently space, frequency, and time are utilized. Expressed in channels/MHz/km^2, channel efficiency is given by[2]

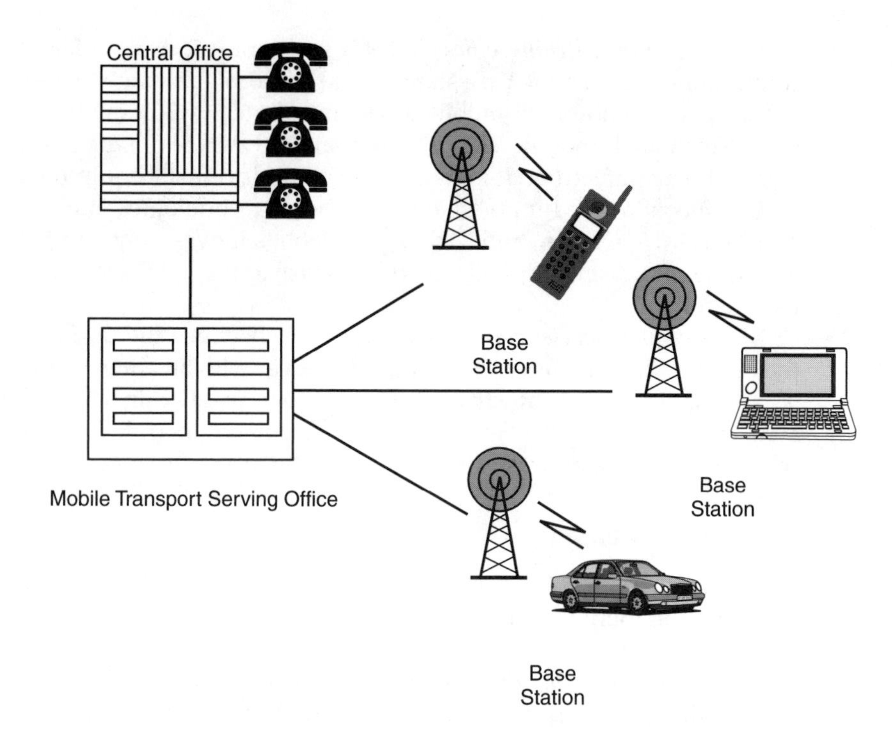

Figure 7.3 A typical cellular network.

$$\eta = \frac{\text{Total no. of channels available in the system}}{\text{Bandwidth} \times \text{Total coverage area}}$$

$$\eta = \frac{\dfrac{B_w}{B_c} \times \dfrac{N_c}{N}}{B_w \times N_c \times A_c} = \frac{1}{B_c \times N \times A_c} \tag{7.1}$$

where B_w is bandwidth of the system in MHz, B_c is the channel spacing in MHz, N_c is the number of cells in a cluster, N is the frequency reuse factor of the system, and A_c is area covered by a cell in km^2.

7.1.2 Cellular network

A typical cellular network is shown in Figure 7.3. It consists of the following three major hardware components:[3]

- *Cell site* (*base stations*) acts as the user-to-mobile telephone switching office interface, as shown in Figure 7.3. It consists of a transmitter and two receivers per channel, an antenna, controller, and data links to the cellular office. Up to 12 channels can operate within a cell, depending on the coverage area.

- *Mobile telephone switching office* (MTSO) is the physical provider of connections between the base stations and the local exchange carrier. MTSO is also known as mobile switching center (MSC) or digital multiplex switch-mobile telephone exchange (DMS-MTX), depending on the manufacturer. It manages and controls cell-site equipment and connections. It supports multiple-access technologies such as AMPS, TDMA, CDMA, and CDPD. As a mobile moves from one cell to another, it must continually send messages to the MTSO to verify its location.
- *Cellular (mobile) handset* provides the interface between the user and the cellular system. It is essentially a transceiver with an antenna and is capable of tuning to all channels (666 frequencies) within a service area. It also has a handset and a number assignment module (NAM), which is a unique address given to each cellular phone.

7.1.3 Cellular standards

The rapid development of cellular technology has resulted in different standards, including the following five:

- *AMPS* (*advanced mobile phone system*) is the standard introduced in 1979. Although it was developed and used in North America, it has also been used in more than 72 countries. It operates in the 800-MHz frequency band and is based on FDMA. As illustrated in Figure 7.4, the mobile transmit channels are in the 825–845 MHz range, while the mobile receive channels are in the 870–890 MHz range. There is also digital AMPS, also known as TDMA (or IS-54). FDMA systems allow for a single mobile telephone to call on a radio channel; each voice channel can communicate with only one mobile telephone at a time. TDMA systems allow several mobile telephones to communicate at the same time on a single radio carrier frequency. This is achieved by dividing their signal into time slots.

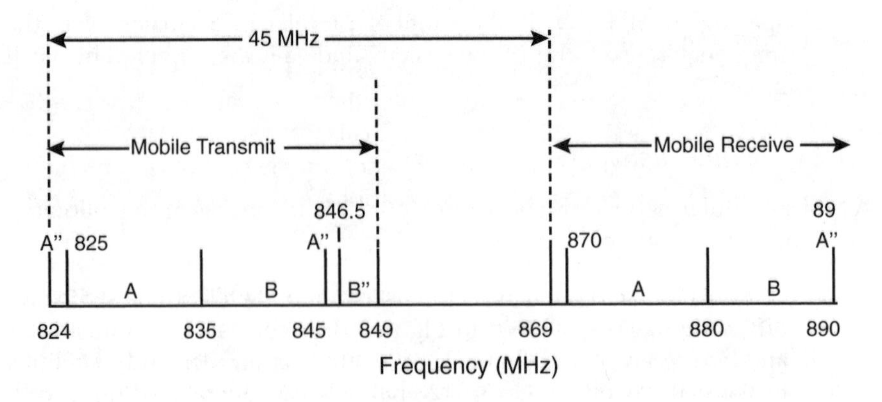

Figure 7.4 FCC cellular frequency allocation.

- *CDMA* (*code-division multiple access*) is an alternative North American cellular standard (IS-95) introduced in 1994. It is based on the spread-spectrum technique, which allows many users to access the same band by assigning a unique orthogonal code to each user.
- *GSM* (*global system for mobile communications*) is a digital cellular standard developed in Europe and designed to operate in the 900-MHz band. It is a globally accepted standard for digital cellular communication. It utilizes a 200-kHz channel divided into eight time slots with FDM. The technology allows international roaming and provides integrated cellular systems across different national borders. GSM is the most successful digital cellular system in the world. It is estimated that many countries outside Europe will join the GSM partnership.
- *PDC* (*personal digital cellular*) is a digital cellular standard developed in Japan. It was designed to operate in the 800-MHz and 1.5-GHz bands.
- *NMT* (*Nordic mobile telephone*) is the cellular standard developed in the Nordic nations of Denmark, Finland, Norway, and Sweden in 1981 and now deployed in over 40 countries in Europe, Asia, and Australia. It was designed to operate in the 450- and 900-MHz frequency bands. NMT 450 and NMT 900 systems can coexist, allowing them to use the same switching center.

In many European countries, the use of GSM has allowed cross-country roaming. However, global roaming has not been realized because there are too many incompatible standards.

7.2 Personal communications systems

The GSM digital network is pervasive in Europe and Asia. The comparable technology PCS is beginning to make inroads in the U.S. PCS is an advanced phone service that combines the freedom and convenience of wireless communications with the reliability of the legacy telephone service. Both GSM and PCS promise clear transmissions, digital capabilities, and sophisticated encryption algorithms to prevent eavesdropping.

PCS is a new concept that will expand the horizon of wireless communications beyond the limitations of current cellular systems to provide users with the means to communicate with anyone, anywhere, anytime. It is called PCS by the FCC but personal communications networks (PCN) by the rest of the world. Its goal is to provide integrated communications (such as voice, data, and video) between nomadic subscribers irrespective of time, location, and mobility patterns. It promises near-universal access to mobile telephony, messaging, paging, and data transfer.

Personal communications enables the internetworking of mobile and fixed network capabilities to allow location/terminal-independent access to services with seamless operation. This is achieved by:[4]

Table 7.1 Comparison of Cellular and PCS Technologies

Cellular	PCS
Fewer sites required to provide coverage	More sites required to provide coverage (e.g., a 20:1 ratio)
More expensive equipment	Less expensive cells
Higher costs for airtime	Airtime costs dropping rapidly
High antenna and more space needed for site	Smaller space for the microcell
Higher power output	Lower power output

- Personal mobility — permitting users to access subscribed services, originate/receive calls at any location on any terminal
- Terminal mobility — enabling mobile terminals at any location to be located and identified by the network
- Service portability — allowing management of user service profiles

PCS/PCN networks and the existing cellular networks should be regarded as complementary rather than competitive. One can view PCS as an extension of cellular to the 1900 MHz band using identical standards. Major factors that separate cellular networks from PCS networks are speech quality, complexity, flexibility of radio-link architecture, economics of serving high-user-density or low-user-density areas, and power consumption of the handsets. Table 7.1 summarizes the differences between the two technologies and services.[2]

PCS offers a number of advantages over traditional cellular communications:

- A truly personal service, combining lightweight phones with advanced features such as paging and voice mail that can be tailored to each individual customer
- Less background noise and fewer dropped calls
- Affordable fully integrated voice and text messaging that works just about anywhere, anytime
- A more secure all-digital network that minimizes chances of eavesdropping or number cloning
- An advanced radio network that uses smaller cell sites
- State-of-the-art billing and operational support system

7.2.1 *Basic features*

PCS refers to digital wireless communications and services operating at broadband (1900 MHz) or narrow-band (900 MHz) frequencies. Thus there are three categories of PCS: broadband, narrow-band, and unlicensed. Broadband PCS addresses both cellular and cordless handset services, while narrow-band PCS focuses on enhanced paging functions. Unlicensed service is

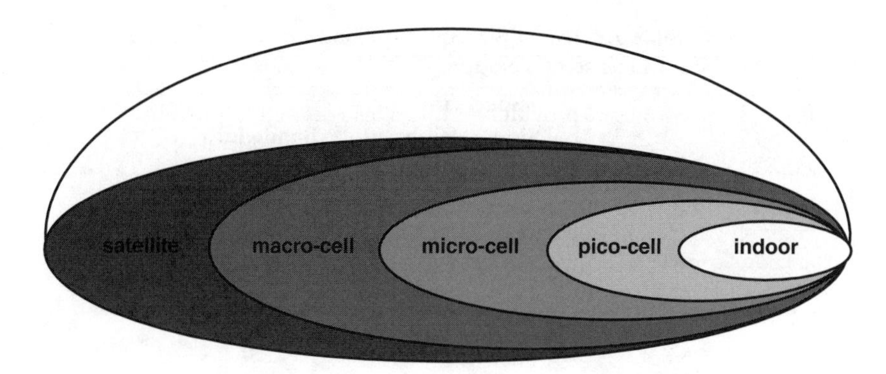

Figure 7.5 Various cell sizes.

allocated from 1890 to 1930 MHz and is designed to allow unlicensed short-distance operation.

The salient features that enable PCS to provide communications with anyone, anywhere, anytime include:[5]

- *Roaming ability:* The roaming service should be greatly expanded to provide universal accessibility.
- *Diverse environment:* Users must be able to use PCS in all types of environment, e.g., urban, rural, commercial, residential, mountains, and recreational areas.
- *Various cell size:* With PCS, there will be a mix of broad types of cell sizes: the picocell for low power indoor applications, the microcell for lower power outdoor pedestrian application; macrocell for high power vehicular applications; and supermacro cell with satellites, as shown in Figure 7.5. For example, a mirocell of a PCS has a radius of 1 to 300 m.
- *Portable handset:* PCS provides a low-power radio, switched access connection to the public switched telephone network (PSTN). The user should be able to carry the handset outside without having to recharge its battery.
- *FCC frequency allocation:* The FCC frequency allocation for PCS usage is significant. FCC allocated 120 MHz for licensed operation and another 20 MHz for unlicensed operation, amounting to a total of 140 MHz for PCS, which is three times the spectrum currently allocated for cellular network. The FCC's frequency allocation for PCS is shown in Tables 7.2 and 7.3 for licensed and unlicensed operators. This is also illustrated in Figure 7.6, where MTA and BTA denote major trading areas and basic trading areas, respectively. We notice that all MTAs are allocated a 15 MHz band, while all BTAs are allocated 5-MHz bands. Six different bands are assigned for MTAs and BTAs.

Table 7.2 The PCS Frequency Bands for Licensed Operation

	Spectrum		
Block	Low Side (MHz)	High Side (MHz)	Bandwidth (MHz)
A	1850–1865	1930–1945	30
D	1865–1870	1945–1950	10
B	1870–1885	1950–1965	30
E	1885–1890	1965–1970	10
F	1890–1895	1970–1975	10
C	1895–1910	1975–1990	30
Total			120

Table 7.3 The PCS Frequency Bands for Unlicensed Operation

Block	Spectrum	Bandwidth (MHz)
Isochronous	1910–1920	10
Asynchronous	1920–1930	10
Total		20

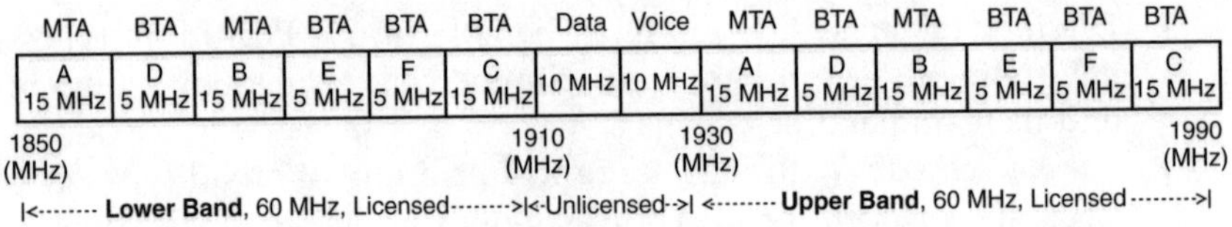

MTA BTA MTA BTA BTA BTA Data Voice MTA BTA MTA BTA BTA BTA
A 15 MHz | D 5 MHz | B 15 MHz | E 5 MHz | F 5 MHz | C 15 MHz | 10 MHz | 10 MHz | A 15 MHz | D 5 MHz | B 15 MHz | E 5 MHz | F 5 MHz | C 15 MHz
1850 (MHz) 1910 (MHz) 1930 (MHz) 1990 (MHz)
|<-------- **Lower Band**, 60 MHz, Licensed-------->|<-Unlicensed->| <-------- **Upper Band**, 60 MHz, Licensed -------->|

Figure 7.6 North American PCS frequency bands.

To use the PCS licensed frequency band, a company must obtain a license from the FCC. To use the unlicensed (or unregulated) PCS spectrum, a company must use equipment that will conform to the FCC unlicensed requirements, which include low power transmission to prevent interference with other users in the same frequency band.

7.2.2 *PCS architecture*

A PCS network is a wireless network that provides communication services to PCS subscribers. The service area of the PCS network is populated with base stations. The base stations are connected to a fixed wire-line network through mobile switch centers (MSCs). Like a cellular network, the radio coverage of a base station is called a cell. The base station locates a subscriber or mobile unit and delivers calls to and from the mobile unit by means of paging within the cell it serves.

PCS architecture resembles that of a cellular network with some differences. The structure of the local portion of a PCS network in shown in Figure 7.7. It basically consists of five major components:[6]

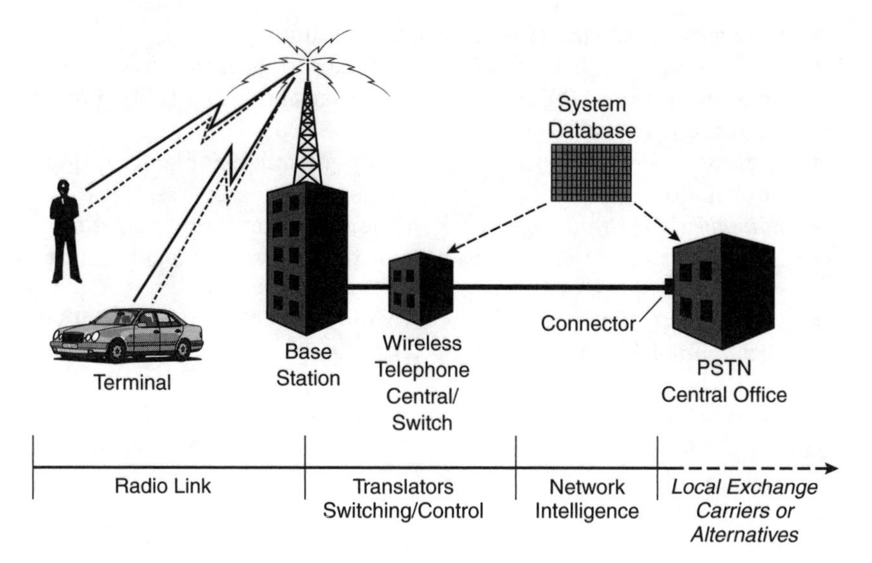

Figure 7.7 Structure of PCS network.

- Terminals installed in the mobile unit or carried by pedestrians
- Cellular base stations to relay signals
- Wireless switching offices that handle switching and routing calls
- Connections to PSTN central office
- Database of customers and other network-related information

Since the goal of PCS is to provide anytime-anywhere communication, the end device must be portable and both real-time interactive communication (e.g., voice) and data services must be available.

7.2.3 *PCS standards*

The Joint Technical Committee (JTC) has been responsible for developing standards for PCS in the U.S. It worked cooperatively with the TIA committee working on the TR-46 reference model and ATIS committee working on the T1P1 reference model.

The TR-46 reference model has the following models:[2]

- *Personal station* (*PS*): Can be a stand-alone device such as a PC or fax machine. PS terminates the radio path on the user side and allows the user to gain access to the network.
- *Radio system* (*RS*): Often known as the base station. It terminates the radio path and connects to the personal communications switching center (PCSC).
- *Personal communication switching center* (*PCSC*): Interfaces the user traffic from the wireless network to the wire-line network.
- *Home location register* (*HLR*): Manages mobile subscribers. It may be part of PCSC or separate.

- *Data message handler* (*DMH*): Used for billing.
- *Visited location register* (*VLR*): Stores subscriber information such as directory number (DN) and electronic serial number (ESN) obtained from the user's HLR.
- *Authentication center*: Manages authentication or encryption of information.
- *Equipment identity register* (*EIR*): Provides information about the PS.
- *Operation system* (*OS*): Takes care of the overall management of the wireless network.
- *Interworking function* (*IWF*): Interconnects PCSC to other networks and enables it to communicate with them.

Some interfaces are specified between the various elements of the system. Based on the TR-46 reference model, a call flows from a PS to an RS to the PCS switch to other network elements. The T1P1 PCS reference model is similar to the TR-46 model but with some differences.

With the TR-46 and T1P1 reference models, PCS has enough capabilities to support a wide range of services. These services include basic services, such as call origination, call termination, call clearing, hand-off, and emergency calls, and supplementary services such as automatic recall, automatic reverse charging, call hold and retrieve, call forwarding-default, call forwarding-busy, call waiting, call transfer, do not disturb, three-way calling, paging, e-mail access and voice message retrieval. It is evident that some of these services are similar to those of the wire-line network or cellular network.

Satellites are instrumental in achieving global coverage and providing PCS services. Mobile satellite communications for commercial users is evolving rapidly toward PCS systems to provide basic telephone, fax, and data services virtually anywhere on the globe. Satellite orbits are being moved closer to the earth, improving communication speed and enabling PCS services. Global satellite systems are being built for personal communications. In the U.S., the FCC licensed five such systems: Iridium, Globalstar, Odyssey, Ellipso, and Aries. In Europe, ICO-Global is building ICO. Japan, Australia, Mexico, and India are making similar efforts.[7] The use of satellite systems for personal communication is discussed further in Chapter 8.

Unlike GSM, PCS is unfortunately not a single standard but a mosaic consisting of several incompatible versions coexisting rather uneasily with one another. One major obstacle to PCS adoption in the U.S. has been the industry's failure to convince customers sufficiently of the advantages of PCS over AMPS, which already offers a single standard. This places the onus on manufacturers to inundate phones with features that attract market attention without compromising the benefits inherent in cellular phones. However, digital cellular technology enjoys distinct advantages. Perhaps the most significant advantage involves security because one cannot adequately encrypt AMPS signals.

There are four major problems currently facing PCS spectrum holders:

- Numerous obstacles to deploying the microcell infrastructure, which requires more towers than cellular.
- Snags in voluntary surrender of the spectrum by current license holders. (Up-front auction costs of $7.7 billion for spectrum licenses make license holders slow to surrender them because they see gold. FCC did recognize this problem and moved to correct it by changing the rule.)
- Marketing, particularly being able to convince users that PCS is better than cellular.
- Economics, particularly PCS costs compared with cellular.

Future growth and success of PCS services cannot be taken for granted. Like any new technology, the success of PCS will depend on a number of factors. These include initial system overall cost, quality and convenience of the service provided, and cost to subscribers. More about PCS can be found in References 8 and 9.

7.3 Cellular digital packet data

Cellular digital packet data (CDPD) is the latest in wireless data communication and offers one of the most advanced means of wireless data transmission technology. CDPD is a cellular standard aimed at providing IP data service over the existing cellular voice networks and circuit switched telephone networks. The technology enables business individuals on the move to communicate data between their work bases and remote locations.

The idea of CDPD was formed in 1992 by a development consortium with key industry leaders, including IBM, six of the seven regional Bell operating companies, and McCaw Cellular. The goal was to create a uniform standard for sending data over existing cellular telephone channels. The Wireless Data Forum (www.wirelessdata.org), formerly known as CDPD Forum, has emerged as a trade association for wireless data service providers, and it currently has over 90 members. CDPD has been defined by the CDPD Forum CDPD Specification R1.1 and operates over AMPS.

By building CDPD as an overlay to the existing cellular infrastructure and using the same frequencies as cellular voice, carriers are able to minimize capital expenditures. It costs approximately $1 million to implement a new cellular cell site and only about $50,000 to build the CDPD overlay to an existing site.

CDPD is designed to exploit the capabilities of the AMPS infrastructure throughout North America. One weakness of cellular telephone channels is that there are moments when the channels are idle (roughly 30% of the air time is unused). CDPD exploits this by detecting and utilizing the otherwise wasted moments by sending packets during the idle time. As a result, data is transmitted without affecting voice system capability. CDPD transmits digital packet data at 19.2 kbps during idle times between cellular voice calls on the cellular telephone network.

CDPD has the following features:

- It is an advanced form of radio communication operating in the 800 and 900 MHz bands.
- It shares the use of the AMPS radio equipment on the cell site.
- It supports multiple, connectionless sessions.
- Its air-link transmission has a 19.2 kbps raw data rate.
- It utilizes IP and the OSI connectionless network protocol (CLNP).
- It is fairly painless for users to adopt. To gain access to CDPD infrastructure, one requires only a special CDPD modem.
- It supports the TCP/IP protocols as well as the international set of equivalent standards.
- It was designed with security in mind, unlike other wireless services — provides for encryption of the user's data as well as conceals the user's identity over the air link.

CDPD provides the following services:

- Data rate of 19.2 kbps.
- Connectionless as the basic service; a user can build a connection-oriented service on top of that if desired.
- All three modes — point-to-point, multicast, and broadcast — are available.
- Security, which involves authentication of users and data encryption.

7.3.1 Network architecture

CDPD is a packet-switched data transfer technology that employs radio frequency (RF) in existing analog mobile phone system such as AMPS. The CDPD overlay network is made up of some major components that operate together to provide the overall service. The following key components define CDPD infrastructure and are illustrated in Figure 7.8:[10,11]

- *Mobile end system* (*MES*): This is the subscriber's device for gaining access to the wireless communication services offered by a CDPD service. It is any mobile computing device that is equipped with a CDPD modem. Examples of an MES are laptop computers, palmtop computers, PDAs, or any portable computing devices. The MES transmits data over the air link to the mobile data base station (MDBS) located in the cell site. The MES is responsible for monitoring the received signal strength of the cellular channel and deciding to initiate a transfer or hand-off from one cell to another cell.
- *Fixed end system* (*FES*): This is a stationary computing device (e.g., a host computer, UNIX workstation, etc.) connected to land-line networks. The FES is the final destination of the message sent from an MES.

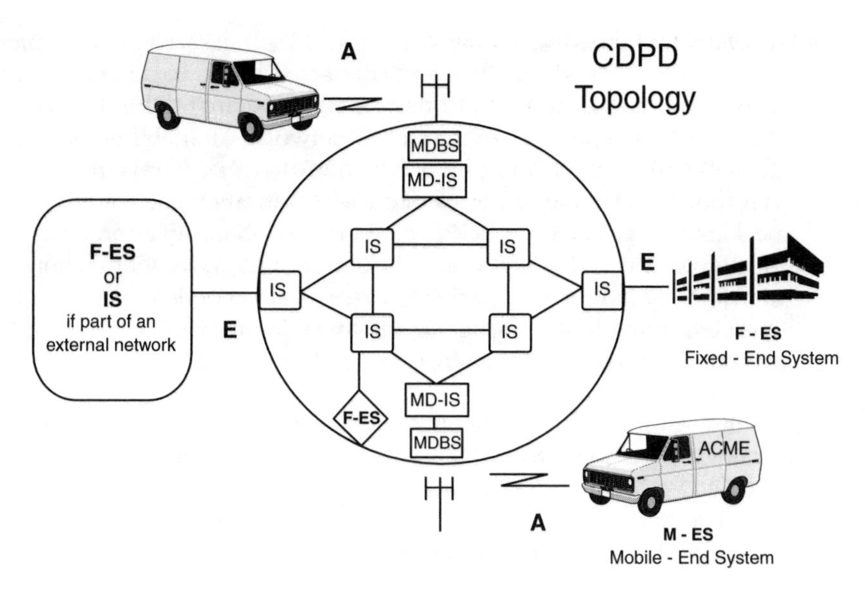

Figure 7.8 Major components of a CDPD network.

- *Intermediate system* (*IS*): This is made up of routers that are CDPD compatible. It is responsible for routing data packets into and out of the CDPD service provider network. It may also perform gateway and protocol conversion functions to aid network interconnection.
- *Mobile data base station* (*MDBS*): CDPD uses a packet-switched system that splits data into small packets and sends them across the voice channel. This involves detecting idle time on the voice channel and sending the packets on the appropriate unoccupied voice frequencies. This detection of unoccupied frequencies and the sending of packets is done by the MDBS. Thus, the MDBS is responsible for relaying data between the mobile units and the telephone network. In other words, it relays data packets from the MES to the MDIS and vice versa. It is responsible for RF channel management. It is housed at the AMPS cell site. It uses the same antenna as the existing cellular network to receive radio signals from the mobile unit and turns them into digital data. The MDBS provides for bidirectional communication for each mobile unit using the reverse and forward channels, and it can communicate with up to 16 mobile units in a sector. Forward channels are contentionless while the reverse channels are shared. The mobile units access the reverse channel using the slotted digital sense multiple access with collision detection (DSMA/CD) protocol similar to the CSMA/CD protocol used in Ethernet. With DSMA/CD, packets are transmitted when the channel is free. Otherwise, the packets are rescheduled for transmission at some random time later as if collision had occurred.

- *Mobile data intermediate system* (*MDIS*): MDBSs that service a particular cell can be grouped together and connected to the backbone router, also known as the mobile data intermediate system (MDIS). The MDIS units form the backbone of the CDPD network. All mobility management functions are taken care of by MDIS. In other words, the MDIS is responsible for keeping track of the MES's location and routing data packets to and from the CDPD network and the MES appropriately. Authorization, following authentication, is also an MDIS function. If an MDIS receives a data packet addressed to a mobile unit within its domain, it sends the packet to the appropriate MDBS. If the data packet is not within its domain, it forwards it to the appropriate MDIS.

Very little new equipment is needed for CDPD service since existing cellular networks are utilized. Only the MDBSs are to be added to each cell. One can purchase CDPD cellular communication systems for Windows or MS-DOS computers. The hardware can be a hand-held AMPS telephone or a small modem that can be attached to a notebook computer. One would need to put up the antenna on the modem.

In order to integrate voice and data traffic effectively on the same cellular network without degrading the service provided for the voice customer, the CDPD network employs a technique known as *channel hopping*. When a mobile unit wants to transmit, it checks for an available cellular channel. Once a channel is found, the data link is established, and the mobile unit can use the assigned channel to transmit as long as the channel is not needed for voice communication. Because voice is king, data packets are sent after voice traffic, which has priority. Therefore, if a cellular voice customer needs the channel, it will take priority over the data transmission. In that case, the mobile unit is advised by the MDBS to "hop" to another available channel. If there are no other available channels, then extra frequencies purposely set aside for CDPD can be used. This situation is rare because each cell typically has 57 channels, and each channel has an average idle time of 25–30%. This process of establishing and releasing channel links is called channel hopping, and it is completely transparent to the mobile data unit. Channel hopping ensures that the data transmission does not interfere with the voice transmission. It usually occurs within the call setup phase of the voice call. Its major disadvantage is the potential interference to the cellular system.

CDPD has been referred to as an "open" technology because it is based on the Open Systems Interconnection (OSI) reference model, as shown in Figure 7.9. The CDPD network is comprised of many layers: layer 1 is the physical layer; layer 2 is the data link layer; and layer 3 is the network layer; etc. For example, the physical layer corresponds to a functional entity that accepts a sequence of bits from the medium access control (MAC) layer and transforms them into a modulated waveform for transmission onto a physical 30 kHz RF channel. The network can use either the ISO connectionless network protocol (CLNP) or the transmission control protocol/Internet protocol (TCP/IP).

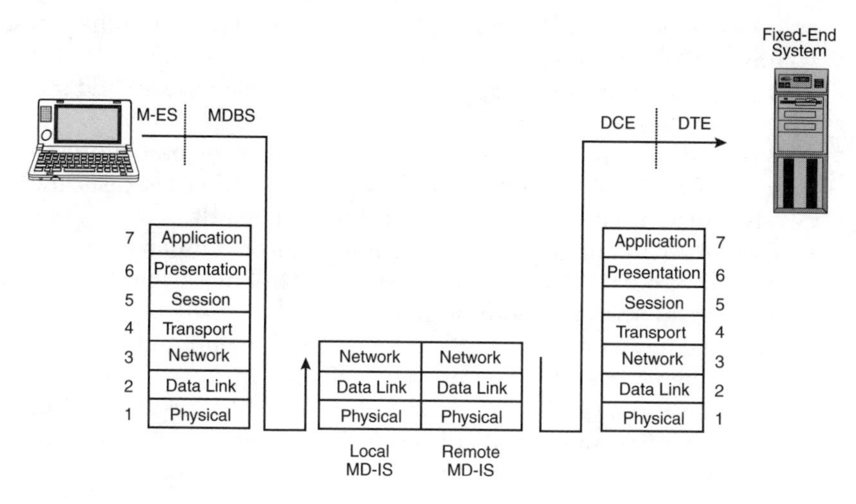

Figure 7.9 OSI reference model.

7.3.2 Applications

CDPD has the following advantages. First, it points to a truly mobile computing environment. Second, by using idle time in voice signals, the cost of data transmission is kept low. Also, cost is based on the volume of data transferred, not on the connection time. Third, because CDPD is an open, nonproprietary standard, it promotes broad availability of hardware and software. It can be used with the existing cellular networks around the world.

CDPD has three major disadvantages. First, its signal is transmitted at roughly 19.2 kbps so effective throughput of the system is only about 9.6 kbps due to the addressing data required for each data packet and the amount of data required to keep the system reliability at a respectable level. The CDPD encodes each block using a systematic Reed-Solomon forward-error-correcting code. Second, redundant MDBS may be needed in each cell in order to improve network reliability. Third, since CDPD is overlaid on the cellular systems, its design is subject to the constraint that no changes be made to the existing cellular systems.

As an integrated technology, CDPD can be applied in a variety of ways. But CDPD is applicable mostly for short bursty-type data applications, not for large file transfers. Generally used as a tool for businesses, CDPD holds promise for improving law enforcement communications and operations. The major applications include the following:[3]

- *Transaction applications*: These applications include credit card verification and charges, dispatches, insurance claims, lottery dispensing machines, ATM machines, remote loan application software, fleet management, package pickup, delivery, and tracking, electronic mail notification and delivery, telemetry, field sales and services, and information retrieval services. Encryption would ensure that these transactions are safe.

- *Batch applications*: These applications include file transfer and statistical information transfer.
- *Broadcast applications*: These applications include general information services (local, state, or international news), weather forecasts, traffic advisories, advertising, and private bulletin boards.
- *Transportation*: Messages can be sent to taxis or company trucks.
- *Internet access*: CDPD provides access to the Internet since both CDPD and the Internet are based on IP.
- *Law enforcement applications*: CDPD has the potential to help police departments operate in a far more efficient manner. An officer on shift can use a CDPD-equipped car on his or her regular patrol to enhance officer safety and efficiency. The officer, linked to headquarters using CDPD, can perform record checks, transmit reports, and send messages to other officers without worrying about compromising privacy or security. The CDPD technology is moving police organizations toward being "paperless," which could significantly increase the operating efficiency of a department covering a large jurisdiction with few officers.

For these applications, CDPD advantages include automated work order routing, transaction documentation, fraud reduction, and improved customer service. As newer technologies evolve, CDPD could be leveraged against those technologies to support wireless data communications. For now, CDPD can coexist with PCS- and CDMA-based infrastructure. More information on CDPD can be found in References 12 and 13.

Summary

Cellular radio was a logical step in the quest for providing additional radio capacity for a geographical area. Cellular systems operate on the principles of cell, frequency reuse, and hand-off.

PCS can be regarded as an extension of the cellular network. It is a communication network where users and the terminals are mobile rather than fixed. The entire complex topic is not yet crystal clear as is typical of such a massive paradigm shift, potentially affecting everyone. Experts predict that PCS will become the primary means of communication for millions of people in North America, while GSM will play the same role throughout Europe and Asia.

CDPD is an overlay to the existing D-AMPS cellular network, which enables users to transmit small chunks of data over a cellular network in a reliable manner, using a portable computing device and a CDPD modem. It allows the mobile user to transmit fax, data, and voice all via one device. CDPD has been implemented using IP as its backbone routing protocol with elements of ISO CLNP used in the mobility management protocols.

References

1. S. Faruque, *Cellular Mobile Systems Engineering*, Artech House, Norwood, MA, 1996, 1–16.
2. V. K. Garg and J. E. Wilkes, *Wireless and Personal Communications Systems*, Prentice-Hall, Upper Saddle River, NJ, 1996, 33–39, 159–225.
3. R. J. Bates, *Wireless Networked Communications*, McGraw-Hill, New York, 1993, 73–95, 129.
4. J. V. Evans, Satellite systems for personal communications, *Proc. IEEE*, vol. 86, no. 7, July 1998, 1325–1341.
5. K. Park, *Personal and Wireless Communications: Digital Technology and Standards*, Kluwer Academic Publishers, Boston, 1996, 1–8.
6. F. J. Ricci, *Personal Communications Systems Applications*, Prentice-Hall, Upper Saddle River, NJ, 1997, 1–14.
7. D. Grillo et al., Guest editorial: Personal communications — services, architecture, and performance issues, *IEEE J. Selected Areas Commun.*, vol. 15, no. 8, Oct. 1997, 1385–1389.
8. Special issue on wireless networks for mobile and personal communications, *Proc. IEEE*, vol. 82, no. 9, Sept. 1994.
9. Special issue on cellular-based personal communications services, *IEEE Personal Commun.*, vol. 4, no. 3, June 1997.
10. N. J. Muller, *Mobile Telecommunications Factbook*, McGraw-Hill, New York, 1998, 303–334.
11. A. K. Salkintzis, Packet data over cellular networks: The CDPD approach, *IEEE Commun. Mag.*, June 1999, 152–159.
12. M. Sreetharan and R. Kumar, *Cellular Digital Packet Data*, Artech House, Norwood, MA, 1996.
13. J. Agosta and T. Russel, *CDPD: Cellular Digital Packet Data Standards and Technology*, McGraw-Hill, New York, 1997.

Problems

7.1 Explain the concepts of cell splitting, frequency reuse, and hand-off in a cellular communications system.

7.2 A cellular system has the following parameters: channel spacing is 50 kHz, the radius of each cell is 2 km, and the frequency reuse factor is 4. Calculate its spectral efficiency.

7.3 Describe the two digital standards developed in North America for cellular systems.

7.4 What is GSM? Why is the dream of global roaming not yet realized?

7.5 Explain the three basic types of cellular technologies: FDMA, TDMA, and CDMA.

7.6 How is PCS/PCN different from cellular technologies?

7.7 In a PCS network, describe the type of information that will flow (a) from the PS to the RS and (b) from the RS to the network.

7.8 CDPD is a blend of digital data transmission and analog cellular technology. Explain.

7.9 Describe the following CDPD elements: MES, MDBS, and MDIS.

7.10 How does channel hopping work in CDPD?

7.11 What are the services provided by CDPD?

7.12 What are the advantages and disadvantages of a CDPD network?

chapter eight

Satellite communications

Everybody is ignorant, only on different subjects.

— Will Rogers

The ultimate goal of communications systems is to provide pocket-sized, wireless devices that will accommodate voice and data services between any two locations on the globe. Long-haul communications and personal communications are emerging to shape a new service that will achieve that goal. The new service ties together the past (satellite) and future (cellular) concepts to offer instant global connectivity — communications from anywhere to anywhere.

Since the launch of the Early Bird satellite (first commercial communication satellite) by NASA in 1965 which proved the effectiveness of satellite communications, satellites have played an important role in both domestic and international communications networks. They have brought voice, video, and data communications to areas of the world that are not accessible by terrestrial lines. By extending communications to the remotest parts of the world, virtually everyone can be part of the global economy.

Satellite (spacecraft) communications is not a replacement of the existing terrestrial systems but rather an extension of wireless systems. However, satellite communication has the following advantages over terrestrial communications:

- *Coverage*: Satellites can cover a much larger geographical area than traditional ground-based systems can. They have the unique ability to cover the globe.
- *High bandwidth*: A Ka-band (27–40 GHz) can deliver throughput at the rate of gigabits per second.
- *Low cost*: A satellite communications system is relatively inexpensive because there are no cable-laying costs and one satellite covers a large area.

Table 8.1 Advantages and Disadvantages of Satellites[1]

Advantages	Disadvantages
Wide-area coverage	Propagation delay
Easy access to remote sites	Dependency on a remote facility
Costs independent of distance	Less control over transmission
Low error rates	Attenuation due to atmospheric particles (e.g., rain) can be severe at high frequencies
Adaptable to changing network patterns	Continual time-of-use charges
No right-of-way necessary, earth stations located at premises	Reduced transmission during solar Equinox

- *Wireless communication*: Users can enjoy untethered mobile communication anywhere within the satellite coverage area.
- *Simple topology*: Satellite networks have simpler topology which results in more manageable network performance.
- *Broadcast/multicast*: Satellites are naturally attractive for broadcast/multicast applications.
- *Maintenance*: A typical satellite is designed to be unattended, requiring only minimal attention by customer personnel.
- *Immunity*: A satellite system will not suffer from disasters, such as flood, fire, and earthquake, and will therefore be available as an emergency service should terrestrial services be disabled.

Of course, satellite systems do have some disadvantages; these are weighed with their advantages in Table 8.1. Some of the services provided by satellites include fixed satellite service (FSS), mobile satellite service (MSS), broadcasting satellite service (BSS), navigational satellite service, and meteorological satellite service.

This chapter explores the integration of satellites with terrestrial networks to meet the demands of highly mobile communities. After presenting the fundamentals of satellite communications, it discusses three major applications of satellite communications: very small aperture terminals (VSATs) for business applications, fixed satellite service (FSS) that interconnects fixed points, and mobile satellite service (MSS) that employs satellites to extend cellular networks to vehicles, ships, and aircraft.

8.1 Fundamentals

A satellite communications system can be viewed as consisting of two parts: the space and ground segments. The space segment consists of the satellites and all their on-board tracking and control system. The earth segment comprises the earth terminals, their associated equipment, and the links to terrestrial networks.[2]

8.1.1 Types of satellites

There were only 150 satellites in orbit in September 1997. The number is expected to be approximately 1700 by the year 2002. With this increasing trend in the number of satellites, there is a need to categorize them according to the height of their orbit and "footprint" or coverage on the earth's surface. They are classified as follows:[3]

- *Geostationary earth orbit (GEO) satellites*: They are launched into a geostationary or geosynchronous orbit, which is 35,786 km above the equator. (Raising a satellite to such an altitude, however, requires a rocket, so that the achievement of a GEO satellite did not take place until 1963.) A satellite is said to be in geostationary orbit when the space satellite, movement is matched to the rotation of the earth at the equator. A GEO satellite can cover nearly one third of the earth's surface, i.e., it takes three GEO satellites to provide global coverage. Due to their large coverage, GEO satellites are ideal for broadcasting and international communications. Examples of GEO satellite constellations are Spaceway, designed by Boeing Satellite Systems, and Astrolink, designed by Lockheed Martin. Another example is Thuraya, designed by Boeing Satellite Systems to provide mobile satellite services to the Middle East and surrounding areas. There are at least three major objections to GEO satellites.[4] First, there is a relatively long propagation delay (or latency) between the instant a signal is transmitted and when it returns to earth (about 240 ms) because of speed-of-light transmission delay and signal processing delay. This delay might not be a problem if the signal is going only one way. However, for signals such as data and voice, which go in both directions, the delay can cause problems. GEO satellites, therefore, are less attractive for voice communication. Second, there is a lack of coverage at far northern and southern latitudes. This coverage gap is unavoidable because a GEO satellite is below the horizon and cannot provide coverage at latitudes as close to the equator as 45°. Unfortunately, many of the European capitals, including London, Paris, Berlin, Warsaw, and Moscow, are north of this latitude. Third, both the mobile unit and the satellite of a GEO system require high transmit power. Despite these objections, the majority of satellites in operation today are GEO satellites, but that might change in the near future.
- *Middle earth orbit (MEO) satellites*: These satellites orbit the earth at 5000 to 12,000 km. GEO satellites do not provide good coverage for places far north and satellites in inclined elliptical orbits are an alternative. Although the lower orbit reduces propagation delay to only 60 to 140 ms round trip, it takes 12 MEO satellites to cover most of the planet. MEO systems represent a compromise between LEO

(described below) and GEO systems, balancing the advantages and disadvantages of each. (MEO are sometimes referred to as *intermediate circular orbit* or ICO).

- *Low earth orbit (LEO) satellites*: LEO satellites circle the earth at 500 to 3000 km. For example, the Echo satellite circled the earth every 90 min. Global coverage could require as many as 200 LEO satellites. Latency in a LEO system is comparable with terrestrial fiber optics, usually less than a 30-ms round trip. LEO satellites are suitable for PCS. However, LEO systems have a shorter life span than the others (5–8 years compared with 12–15 years for GEO systems) due to the increased amount of radiation in the low-earth orbit. LEO systems have been grouped as Little LEO and Big LEO. Little LEOs have less capacity and are limited to nonvoice services such as data and message transmission. An example is OrbComm, designed by Orbital Corporation, which consists of 36 satellites, each weighing 85 lb. Big LEOs have larger capacity and voice transmission capability. An example is Loral and Qualcomm's Globalstar, which will operate in the L-band frequencies and employ 48 satellites organized in eight planes of six satellites each.

The arrangement of the three basic types of satellites is shown in Figure 8.1. Both MEO and LEO satellites are regarded as non-GEO satellites. As seen from the earth station, the GEO satellite never appears to move any significant distance. As seen from the satellite, the earth station never appears to move. The MEO and LEO satellites, on the other hand, as seen from the ground are continuously moving. Likewise, the earth station, as seen from the satellites, is a moving target. GEO systems are well suited for the delivery of broadcast services, while LEO systems are very efficient for the delivery of interactive services. LEO satellites tend to be smaller, lighter, and cheaper than MEO, which in turn are likely to be less expensive than GEO. The evolution from GEO to MEO and LEO satellites has resulted in a variety of global satellite systems. The convenience of GEO was weighed against the practical difficulty involved with it and the inherent technical advantages of LEO, such as lower delay and higher angles of elevation. While it has been

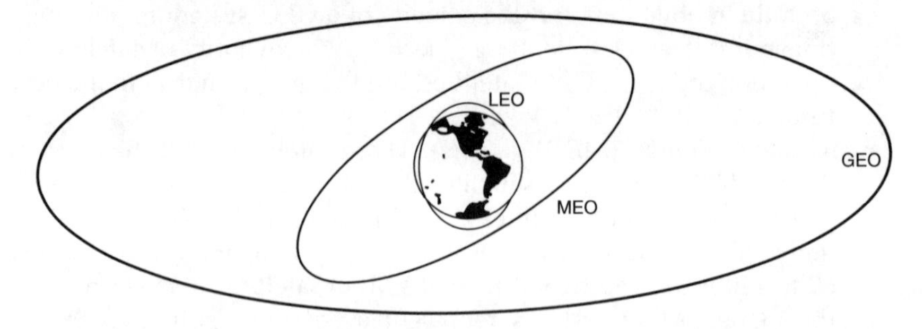

Figure 8.1 The three common types of satellites — GEO, MEO, and LEO.

conceded that GEO is in many respects theoretically preferable, LEO or MEO systems are preferred for many applications. Although a constellation (a group of satellites) is required instead of only one for hemispheric coverage, the loss of individual satellites would cause only gradual degradation of the system rather than a catastrophic failure.

Several attractive services can be offered by utilizing the small satellite technology. These include (1) personal communication service (PCS) in a global scale, (2) digital audio broadcasting (DAB), (3) environmental data collection and distribution, (4) remote sensing/earth observation, and (5) several military applications.

8.1.2 Frequency bands

Every nation of the world has the right to access the satellite orbit and no nation has a permanent right or priority to use any particular orbit location. Without a means for nations to coordinate use of satellite frequency band, the satellite services of one nation could interfere with those of another, thereby creating a chaotic situation in which neither country's signals could be received clearly.

To facilitate satellite communications and eliminate interference between different systems, international standards govern the use of satellite frequency. The International Telecommunication Union (ITU) is responsible for allocating frequencies to satellite services. As far as frequency allocation is concerned, the entire world has been divided into three regions:

- *Region 1*: Europe, Africa, the Middle East, Mongolia, and the Asian region of the former USSR
- *Region 2*: North and South America and Greenland
- *Region 3*: The remainder of Asia, Australia, and the southwest Pacific

Since the spectrum is a limited resource, the ITU has reassigned the same parts of the spectrum to many nations and for many purposes throughout the world.

The frequency spectrum allocations for satellite services are given in Table 8.2. Notice that the assigned segment is the 1–40 GHz frequency range,

Table 8.2 Satellite Frequency Allocations

Frequency Band	Range (GHz)
L	1–2
S	2–4
C	4–8
X	8–12
Ku	12–18
K	18–27
Ka	27–40

Table 8.3 Typical Uplink and Downlink Satellite Frequencies

Uplink Frequencies (GHz)	Downlink Frequencies (GHz)
5.925–6.425	3.700–4.200
7.900–8.400	7.250–7.750
14.00–14.50	11.70–12.20
27.50–30.0	17.70–20.20

which is the microwave portion of the spectrum. As microwaves, the signals between the satellite and the earth stations travel along line-of-sight paths and experience free-space loss that increases as the square of the distance.

Satellite services are classified into 17 categories:[5] fixed, intersatellite, mobile, land mobile, maritime mobile, aeronautical mobile, broadcasting, earth exploration, space research, meteorological, space operation, amateur, radiodetermination, radionavigation, maritime radionavigation, and standard frequency and time signal. The Ku band is presently used for broadcasting services and also for certain fixed satellite services. The C band is exclusively for fixed satellite services, and no broadcasting is allowed. The L band is employed by mobile satellite services and navigation systems.

A satellite band is divided into separation portions: one for earth-to-space links (the uplink) and one for space-to-earth links (the downlink). Separate frequencies are assigned for sending to the satellite (the uplink) and receiving from the satellite (the downlink). Table 8.3 provides the general frequency assignments for uplink and downlink satellite frequencies. As shown in the table, the uplink frequency bands are slightly higher than the corresponding downlink frequency band. The difference is to take advantage of the fact that it is easier to generate RF power within a ground station than it is on a satellite. In order to direct the uplink transmission to a specific satellite, the uplink radio beams are highly focused. In the same way, the downlink transmission is focused on a particular *footprint* or area of coverage.

All satellite systems are constrained to operate in designed frequency bands depending on the kind of earth station used and service provided. The satellite industry, particularly in the U.S., is subject to several regulatory requirements, domestically and internationally, depending upon which radio services and frequency bands are proposed to be used on the satellite. In the U.S., the FCC is the independent regulatory agency that ensures that the limited orbital/spectrum resource allocated to space radiocommunications services is used efficiently. After receiving an application for a U.S. domestic satellite, the FCC initiates the advance publication process for a U.S. satellite. This action is to ensure the availability of an orbit position when the satellite is authorized. The FCC does not guarantee international recognition and protection of satellite systems unless the authorized satellite operator complies with all coordination requirements and completes the necessary coordination of its satellites with all other administrations whose satellites are affected.[6]

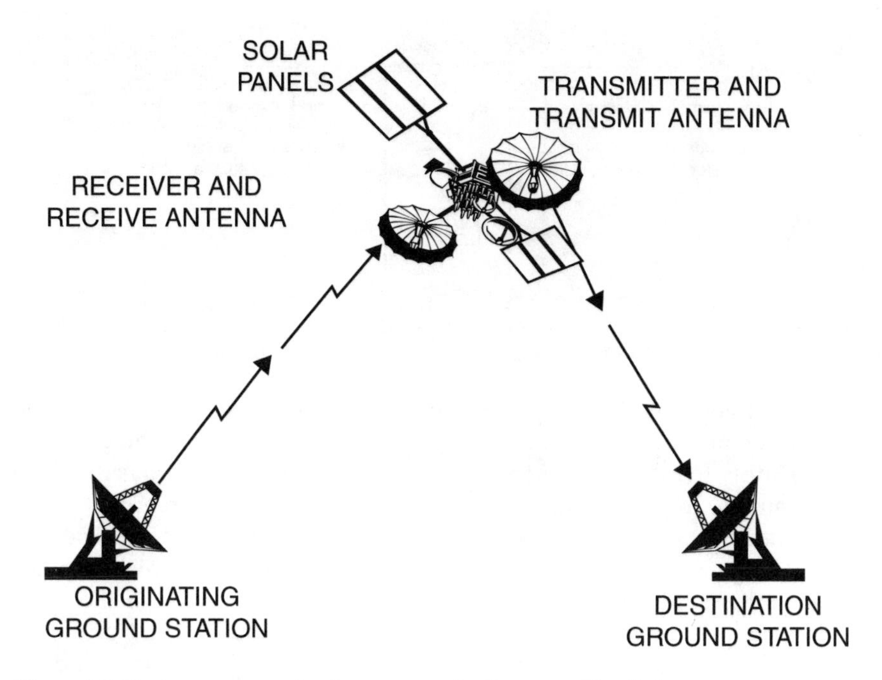

Figure 8.2 Basic components of a communications satellite link.

8.1.3 Basic satellite components

Every satellite communication involves the transmission of information from a ground station to the satellite (the uplink), followed by a retransmission of the information from the satellite back to earth (the downlink). Hence, the satellite must typically have a receiver antenna, receiver, transmitter antenna, transmitter, some mechanism for connecting the uplink with the downlink, and a power source to run the electronic system. These components are illustrated in Figure 8.2 and explained below:

- *Transmitters*: The amount of power a satellite transmitter is required to send out depends on whether it is a GEO or a LEO satellite. GEO satellites are about 100 times farther away than are LEO satellites. Thus, a GEO would need 10,000 times as much power as a LEO satellite. Fortunately, other parameters can be adjusted to reduce this amount of power.
- *Antennas*: The antennas dominate the appearance of a communication satellite. The antenna design is one of the more difficult and challenging parts of a communication satellite project. Antenna geometry is constrained physically by the design and the satellite topology. A major difference between GEO and LEO satellites is their antennas. Since all the receivers are located in the coverage area, which is

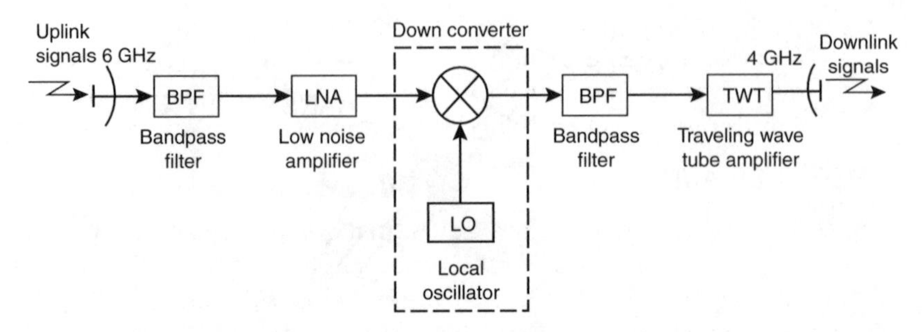

Figure 8.3 A simplified block diagram of a satellite transponder.

relatively small, a properly designed antenna can focus most of the transmitter power within that area The easiest way to achieve this is simply to make the antenna larger which is one of the ways the GEO satellite makes up for the larger transmitter power it requires.

- *Power generation*: The satellite must generate all of its own power. The power is often generated by large solar cells, which convert sunlight into electricity. Since there a limit to how large the solar panel can be, there is also a practical limit to the amount of power that can be generated. Satellites must also be prepared for periods of eclipse, when the earth is between the sun and the satellite. Eclipses necessitate having batteries on board that can supply power during eclipse and recharge later.
- *Transponders*: These are the devices each satellite must carry. They receive radio signals at one frequency, amplify them, and convert them to another for transmission, as shown typically in Figure 8.3. For example, a GEO satellite might have 24 transponders, each assigned a pair of frequencies (uplink and downlink frequencies).

8.1.4 *Effects of space*

Space has two major effects space on satellite communications. First, the space environment, with radiation, rain, and space debris, is hard on satellites. The satellite payload, which is responsible for the satellite communication functions, is expected to be simple and robust. Traditional satellites, specially GEOs, serve as bent pipes and act as repeaters between communication points on the ground. There is no on-board processing (OBP). However, new satellites allow OBP, including decoding/recoding, demodulation/remodulation, transponding, beam switching, and routing.[7]

The second effect is that of wave propagation, discussed in Chapter 5. Line-of-sight and attenuation due to atmospheric particles (rain, ice, dust, snow, fog, etc.) are not significant at L-, S-, and C-bands. Above 10 GHz, the main propagation effects are:[8,9]

- *Tropospheric propagation effects:* attenuation by rain and clouds, scintillation, and depolarization
- *Effects of the environment on mobile terminals:* shadowing, blockage, and multipath caused by objects in the area surrounding the terminal antenna

The troposphere can produce significant signal degradation at the Ku-, Ka-, and V-bands, particularly at lower elevation angles. Most satellite systems are expected to operate at an elevation angle above approximately 20°. Rain constitutes the most fundamental obstacle encountered in the design of satellite communication systems at frequencies above 10 GHz. The resultant loss of signal power makes for unreliable transmission.

8.2 Orbital characteristics

An intuitive question is "What keeps objects in orbit?" The answer is found in the orbital mechanical laws governing satellite motion. Satellite orbits obey the same laws of Johannes Kepler that govern the motion of planets around the sun. Kepler's three laws follow:[10]

- *First law*: The orbit of each planet follows an elliptical path in space with the sun serving as the focus.
- *Second law*: The line linking a planet with the sun sweeps out equal areas in equal time.
- *Third law*: The square of the period of a planet is proportional to the cube of its mean distance from the sun.

Besides these laws, Newton's law of gravitation also plays a part. It states that any two bodies attract each other with a force proportional to the product of their masses and inversely proportional to the square of the distance between them, i.e.,

$$\mathbf{F} = -\frac{GMm}{r^2}\mathbf{a}_r \tag{8.1}$$

where M is the one body (earth), m is the mass of other body (satellite), $\mathbf{F}$ is the force on m due to M, r is the distance between the two bodies, $\mathbf{a}_r = \mathbf{r}/r$ is a unit vector along the displacement vector $\mathbf{r}$, and $G = 6.672 \times 10^{-11}$ Nm/kg^2 is the universal gravitational constant. If M is the mass of the earth, the product $GM = \mu = 3.99 \times 10^{14}$ m^3/s^2 is known as Kepler's constant.

Kepler's laws in conjunction with Newton's laws can be used to describe completely the motion of the planets around the sun or that of the satellite around the earth. Newton's second law can be written as

$$\mathbf{F} = m\frac{d^2r}{dt^2}\mathbf{a}_r \tag{8.2}$$

Equating this with the force between the earth and the satellite in Equation 8.1 gives

$$\frac{d^2r}{dt^2}\mathbf{a}_r = -\frac{\mu}{r^2}\mathbf{a}_r \tag{8.3}$$

or

$$\ddot{\mathbf{r}} + \frac{\mu}{r^3}\mathbf{r} = 0 \tag{8.4}$$

where $\ddot{\mathbf{r}}$ is the vector acceleration. The solution to the vector second-order differential Equation 8.4 is not simple, but it can be shown that the resulting trajectory is in the form of an ellipse given by[10,11]

$$r = \frac{p}{1 + e\cos\theta} \tag{8.5}$$

where r is the distance between the geocenter and any point on the trajectory, p is a geometric constant, e ($0 \le e < 1$) is the eccentricity of the ellipse, and θ (known as the true anomaly) is the polar angle between r and the point on the ellipse nearest to the focus. These orbital parameters are illustrated in Figure 8.4. The point on the orbit where the satellite is closest to the earth is known as the *perigee*, while the point where the satellite is farthest from the earth is known as the *apogee*. The fact that the orbit is an ellipse confirms

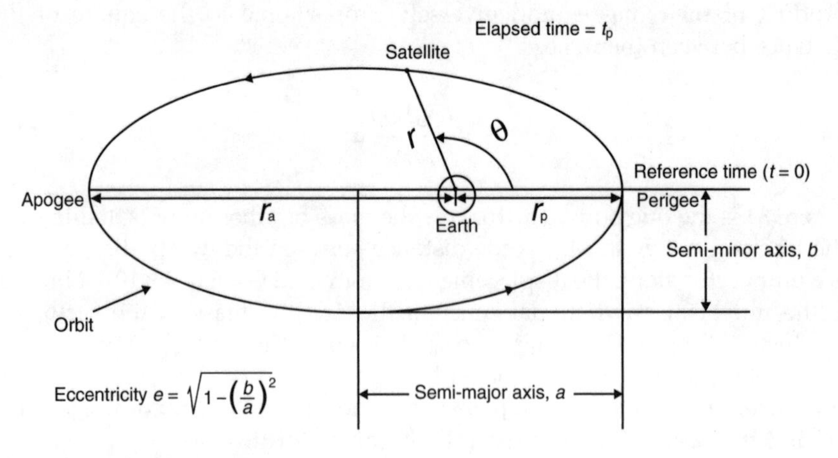

Figure 8.4 Orbital parameters.

Kepler's first law. If a and b are the semimajor and semiminor axes (see Figure 8.4), then

$$b = a\sqrt{(1 - e^2)} \tag{8.6}$$

$$p = a(1 - e^2) \tag{8.7}$$

Thus, the distance between a satellite and the geocenter is given by

$$r = \frac{a(1 - e^2)}{1 + e\cos\theta} \tag{8.8}$$

Note that the orbit becomes circular orbit when $e = 0$.

The apogee height and perigee height are often required. From the geometry of the ellipse, the magnitudes of the radius vectors at apogee and perigee can be obtained as

$$r_a = a(1 + e) \tag{8.9a}$$

$$r_p = a(1 - e) \tag{8.9b}$$

To find the apogee and perigee heights, the radius of the earth must be subtracted from the radii lengths.

The period T of a satellite is related to its semimajor axis a using Kepler's third law as

$$T = 2\pi\sqrt{\frac{a^3}{\mu}} \tag{8.10}$$

For a circular orbit to have a period equal to that of the earth's rotation — a sidereal day: 23 h, 56 min, 4.09 s — an altitude of 35,803 km is required. In this equatorial plane, the satellite is "geostationary."

The velocity of a satellite in an elliptic orbit is obtained as

$$v^2 = \mu\left(\frac{2}{r} - \frac{1}{a}\right) \tag{8.11}$$

For a synchronous orbit (T = 24 h), $r = a = 42{,}230$ km, and $v = 3074$ m/s or 11,070 km/h.

A constellation is a group of satellites. The total number N of satellites in a constellation depends on the earth central angle γ and is given by[12]

$$N \approx \frac{4\sqrt{3}}{9}\left(\frac{\pi}{\gamma}\right)^2 \tag{8.12}$$

To determine the amount of power received on the ground due to satellite transmission, we consider the power density

$$\Psi = \frac{P_t}{S} \tag{8.13}$$

where P_t is the power transmitted and S is the terrestrial area covered by the satellite. The value of P_t is a major requirement of the spacecraft. The coverage area is given by

$$S = 2\pi R^2(1-\cos\gamma) \tag{8.14}$$

where R = 6378 km is the radius of the earth. S is usually divided into a cellular pattern of spot beams, thereby enabling frequency reuse. The effective area of the receiving antenna is a measure of the ability of the antenna to extract energy from the passing electromagnetic wave and is given by

$$A_e = G_r\frac{\lambda^2}{4\pi} \tag{8.15}$$

where G_r is the gain of the receiving antenna and λ is the wavelength. The power received is the product of the power density and the effective area. Thus,

$$P_r = \Psi A_e = \frac{G_r\lambda^2}{4\pi S}P_t \tag{8.16}$$

This is known as the Friis equation, relating the power received by one antenna to the power transmitted by the other. We first notice from this equation that for a given transmitted power P_t, the received power P_r is maximized by minimizing the coverage area S. Second, mobile terminals prefer having nondirectional antennas, thereby making their gain G_r fixed. Therefore, maximizing P_r encourages use of as long a wavelength as possible, i.e., as low a frequency as practicable within regulatory and technical constraints.

The noise density N_o is given by

$$N_o = kT_s \tag{8.17}$$

where $k = 1.38 \times 10^{-23}$ Ws/°K is Boltzmann's constant and T_s is the equivalent system temperature, which is defined to include antenna noise and thermal noise generated at the receiver. Shannon's classical capacity theorem for the maximum error-free transmission rate in bits per second (bps) over a noisy power-limited and bandwidth-limited channel is

$$C = B \log_2\left(1 + \frac{P_r}{N_o}\right) \tag{8.18}$$

where B is the bandwidth and C is the channel capacity.

8.3 VSAT networks

A very small aperture terminal (VSAT) is a dish antenna that receives signals from a satellite. (The dish antenna has a diameter that is typically in the range of 1.2 to 2.8 m, but the trend is toward smaller dishes no more than 1.5 m in diameter.) A VSAT may also be regarded as a complete earth station that can be installed on the user's premises and provide communication services in conjunction with a larger (typically 6–9 m) earth station acting as a network management center (NMC), as illustrated in Figure 8.5.

VSAT networks arose in the mid-1980s as a result of electronic and software innovations that allowed all the required features to be contained in an affordable package about the size of a PC. VSAT technology brings features and benefits of satellite communications down to an economical and usable form. VSAT networks have become mainstream networking solutions

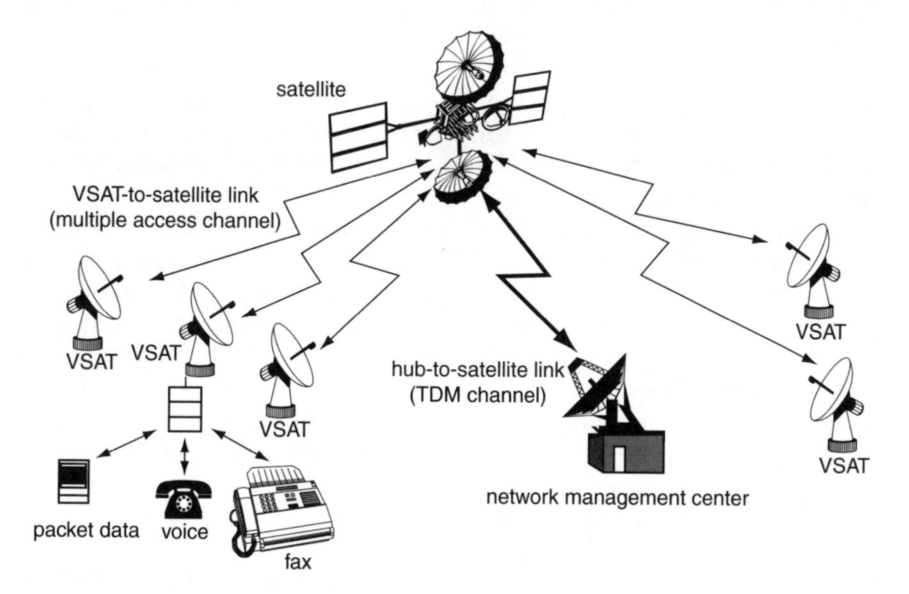

Figure 8.5 A typical VSAT network.

for long-distance, low-density voice and data communications because they are affordable to both small and large companies. Other benefits and advantages of VSAT technology include lower operating costs, ease of installation and maintenance, ability to manage multiple protocols, and ability to bring into the communication loop locations where the cost of leased lines is very high.

8.3.1 Network architecture

Satellite links can support interactive data applications through two types of architectures:[3,13] mesh topology (also called point-to-point connectivity) and star topology (also known as point-to-multipoint connectivity). Single-hop communications between remote VSATs can be achieved by full-mesh connectivity.

The star network employs a hub station, as shown in Figure 8.6a. The hub consists of an RF terminal, a set of baseband equipment, and network equipment. A VSAT network can provide transmission rates up to 64 kbps. As is common with star networks, all communication must past through the hub. That is, all communication is between the remote node and the hub; no direct node-to-node information transfer is allowed in this topology. This type of network is highly coordinated and can be very efficient. The point-to-multipoint architecture is very common in modern satellite data networks and is responsible for the success of the current VSAT. Its simple mode of communications makes it useful for businesses; the hub station is located at the company headquarters while the remote VSATs are located at the branches. Most VSAT star networks employ the TDMA access method for the in-route (from VSAT to hub) and TDM for the out-route (from the hub to the VSAT).

A mesh network is more versatile than a star network because it allows any-to-any communications. Also, the star network can provide transmission rates only to 64 kbps per remote terminal, whereas the mesh network can have its data rates increased to 2 Mbps or more. Although the mesh and star configurations have different technical requirements, it is possible to integrate the two if necessary.

Mesh topology was used by the first satellite networks to be implemented. With time, there was a decline in the use of this topology, but it remains an effective means of transferring information with the least delay. Mesh topology applies to either temporary connections or dedicated links to connect two earth stations. As shown in Figure 8.6b, all full-duplex point-to-point connectivities are possible and provided, as typical of a mesh configuration. If there are N nodes, the number of connections is equal to the permutation of $N(N-1)/2$. Mesh networks are implemented at C and Ku bands. The transmission rate ranges from about 64 kbps to 2.048 Mbps (E1 speed). Users have implemented 45 Mbps.

Transponder resources are assigned to VSATs on the network using transponder access protocols. Three such protocols are FDMA, TDMA, and CDMA. FDMA is the most popular access method because it allows the use

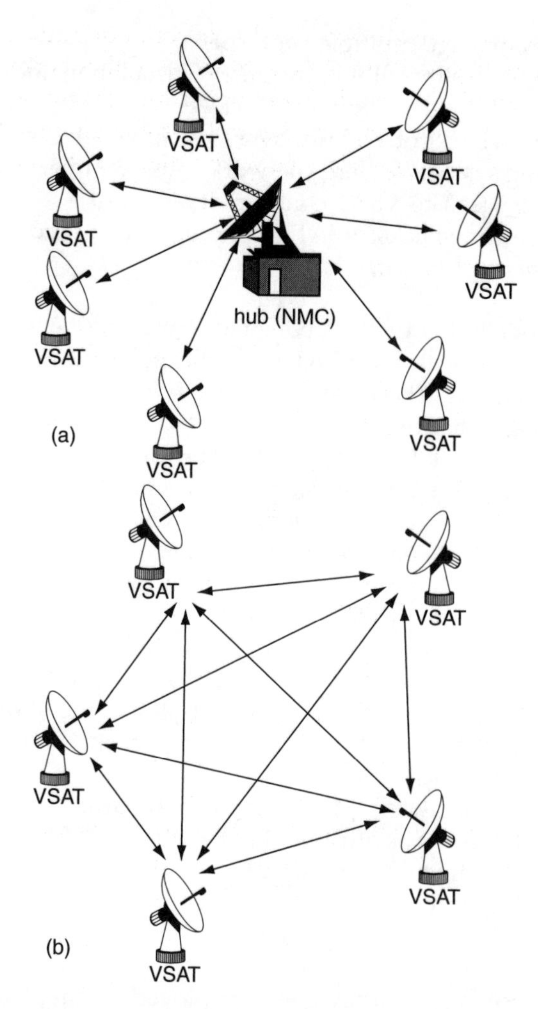

Figure 8.6 VSAT architecture: (a) star network, (b) mesh network.

of comparatively low-power VSAT terminals. TDMA is not efficient for low-density up-link traffic from the VSAT, which is mostly data transfer of a bursty nature.

8.3.2 Applications

Although TV used to dominate domestic satellite communications, data has grown tremendously with the advent of VSAT. VSAT technology has the following advantages:[14]

- Ability to extend communications to remote areas where provision of terrestrial facilities can be very expensive
- Ability to configure or reconfigure a network quickly

- Ability to manage multiple protocols
- Single-vendor support for equipment installation and maintenance

The major drawbacks of VSAT networks are the high costs, the tendency toward optimizing systems for large networks (for example, with 500 VSATs), and lack of direct VSAT-to-VSAT connectivity.

Despite these drawbacks, VSAT technology has found applications in the following areas:[2]

- *PC integration with VSAT*: A personal computer (PC) serves as a direct user interface with the VSAT in applications requiring online information delivery. The most popular service of this kind is DirecPC, an offering of hardware, software, and satellite service delivery from Hughes Network Systems (HNS). The DirecPC package includes a satellite dish and an expansion card designed for a PC's I/O bus.
- *Integrating LANs with VSAT networks*: A VSAT star network can create an efficient WAN environment for interconnection of legacy LANs. The functionality provided is similar to that of LAN-to-LAN interconnection using bridges, routers, and gateways. For example, LANAdvantage is a specific implementation of LAN connectivity with a VSAT star network. It is an HNS product that operates at the MAC sublayer of the link layer and allows the VSAT network to transfer the network layer and higher packet data on a completely transparent basis.
- *Television broadcasting*: Much of the popularity of direct-broadcast satellite (DBS) can be attributed to the quality of TV communications provided by digital technology. For example, DirecTV provides personalized TV through programming capabilities. It was the first high-power DBS service to deliver up to 175 channels of digital-quality programming.
- *Others*: VSATs are mainly being employed as a replacement for terrestrial data networks using analog private lines in various industries, such as teleconferencing, training, retailing, insurance, credit-card checking, reservation systems, interactive inventory data sharing, automobile sales and distribution, banking, travel reservations, lodging, and finance. For example, Walmart employed VSATs to extend its reach to thousands of remote towns in the rural U.S. Chrysler Corporation has provided all of its 6000 U.S. dealers with a VSAT to be used for order entry. Space does not permit mentioning similar applications by GE, Holiday Inn, Toyota of America, etc.

The high transmission rates (typically 155 and 622 Mbps) associated with ATM are well above the maximum rates possible with today's VSAT technology. However, most users require lower traffic rates. These users will need a cost-efficient way to access the BISDN/ATM network and take advantage of the bandwidth-on-demand property of ATM.[15]

Several types of VSAT networks are now in operation, both domestically and internationally. There were over 1000 VSATs in operation at the beginning of 1992. Today, there are over 100,000 two-way Ku-band VSATs installed in the U.S. and over 300,000 worldwide. Almost all of these VSATs are designed primarily to provide data for private corporate networks, and almost all two-way data networks with more than 20 earth stations are based on some variation of an ALOHA protocol for access.[16,17] The price of a VSAT started around $20,000 and dropped to around $6,000 in 1996.

8.4 Fixed satellite service

Several commercial satellite applications are through earth stations at fixed locations on the ground. The international designation for such an arrangement is *fixed satellite service* (FSS). FSS provides communication service between two or more fixed points on earth, as opposed to mobile satellite services (MSS) which provide communication for two moving terminals. (MSS is discussed in the next section.) Although ITU defined FSS as a space radiocommunication service covering all types of satellite transmissions between given fixed points, the line between FSS and broadcasting satellite service (BSS) for satellite television is becoming more and more blurred[18]. FSS applies to systems that interconnect fixed points such as international telephone exchanges. It involves GEO satellites providing 24-hour per day service.

A complete link can be established via the FSS with a variety of configurations, including:[2]

- Direct links such as business-to-business
- Connections via terrestrial links
- Connections to VSATs

Table 8.4 shows the WARC (World Administrative Radio Conference) frequency allocations for FSS. The table gives only a general idea and is not at all comprehensive. FSS shares frequency bands with terrestrial networks in the 6/4 GHz and 14/12 GHz bands. Thus, it is possible that a terrestrial network could affect a satellite on the uplink or that a terrestrial network may be affected by the downlink from a satellite.

Table 8.4 Frequency Allocations for FSS (below ~30 GHz)

Downlinks (in GHz)	Uplinks (in GHz)
3.4–4.2 and 4.5–4.8	5.725–7.075
7.25–7.75	7.9–8.4
10.7–11.7	
11.7–12.2 (Region 2 only)	12.75–13.25 and 14.0–14.5
12.5–12.75 (Region 1 only)	
17.7–21.2	27.5–31.0

As exemplified by Intelsat, FSS has been the most successful part of commercial satellite communications. Early applications were point-to-point telephony and major trunking uses. Current applications of the FSS can be classified according to frequency (from about 3 MHz to above 30 GHz); the HF band is the lowest frequency. The categories include:

- *High frequency service*: The high frequency (HF) bands have been crowded because high frequency service is the only technology that can provide very long-range coverage with a minimum investment in infrastructure. Since HF signals are reflected to earth by the ionosphere, they can travel long distances. Because one can achieve the long-range capability with inexpensive equipment, HF is valuable for many long-range fixed applications. However, HF communication users must consider the constantly changing nature of the ionosphere, high levels of ambient noise, interference, and the need for relatively large antennas. The Department of Defense (DOD) and many other federal agencies use HF fixed service to support priority communications after natural disasters, such as earthquakes and hurricanes, or to maintain many HF links to overseas bases, embassies, and offices. Many private industries use HF links to communicate with their foreign offices.
- *Private fixed services*: These microwave services are licensed by the FCC and include services operated by organizations mainly to carry signals for their own purposes. Major users include private companies, utilities, transportation providers, and local and state governments.
- *Auxiliary broadcasting (AUXBC) services*: AUXBC services include applications that support the TV and AM/FM broadcasting industries. Electronic news gathering (ENG) uses transportable microwave links to provide live coverage of events. AM and FM stations use such services for temporary live coverage of events as well as for studio-to-transmitter links.
- *Cable relay service (CARS):* CARS supports the TV industry. CARS and AUXBC use ENG in identical ways to provide temporary real time coverage of events outside the studio.
- *Federal government fixed services*: The U.S. government uses fixed services for many functions. Federal civilian agencies such as the Department of Interior and Department of Agriculture use microwave networks to support mobile radiocommunication sites in federally controlled remote areas such as National Parks and National Forests. The Departments of Justice and Treasury maintain extensive urban radio networks to support national law enforcement and security. Numerous federal agencies make extensive use of fixed services to communicate with a wide range of sensors that track airways, weather, stream flow, etc. Military operations and training make extensive use of fixed satellite microwave terminals to support range safety and security, relay data received from airborne platforms to central

control sites, to provide closed circuit TV for safety, to provide radar tracking and air traffic control information, and to support a wide range of logistical and administrative support activities. NASA's Advanced Communications Technology Satellite (ACTS) is a major development for FSS. Launched in September 1993, ACTS is designed to verify new FSS technologies.

In Europe, Esa was the catalyst for FSS development, but it was later turned over to the Eutelsat system to form the basis for a European domestic and regional system. The Eutelsat system comprises seven GEO satellites that support telephony, business services, television, radio, and mobile satellite communications.

In Japan, current FSS operations are provided by the CS-3 series of satellites. The N-Star satellite is the next generation FSS satellite being developed by NTT. It includes Ka-band, Ku-band, and C-band transponders for FSS. For earth stations, NTT has built a dual beam antenna using one main reflector.

In Russia, FSS is currently provided by the Gorizont system with transponders at 4/6 GHz and 11/14 GHz. The demand for FSS development in Russia is limited due to the lack of a strong telephone infrastructure. The next step is a system designated as Express which will be used for television transmission.

Although the telecommunications industry as a whole is growing rapidly, the FSS industry is not. The market trend is toward the replacement of long-haul microwave systems with fiber. Fiber provides much greater capacity than does microwave.

8.5 Mobile satellite service

There is a need for global cellular service in all geographical regions of the world. The terrestrial cellular systems serve urban areas well; they are not economical for rural or remote areas where the population or tele-density is low. Mobile satellite (MSAT) systems can complement the existing terrestrial cellular network by extending communication coverage from urban to rural areas. Mobile satellite services (MSS) are not limited to land coverage but include marine and aeronautical services.[5,19] Thus, the coverage of mobile satellites is based on geographical coverage, not on population coverage as in terrestrial cellular systems, and could be global.

MSAT or satellite-based PCS/PCN is being developed in light of the terrestrial constraints. The low cost of installation makes satellite-based PCS simple and practical. The American Mobile Satellite Corporation (AMSC) and Telesat Mobile of Canada are designing a geosynchronous MSAT to provide PCS to North America. The concept of MSAT is illustrated in Figure 8.7.

Satellite communication among mobile earth stations is different from the cellular communication discussed in Chapter 7. (The two cellular systems

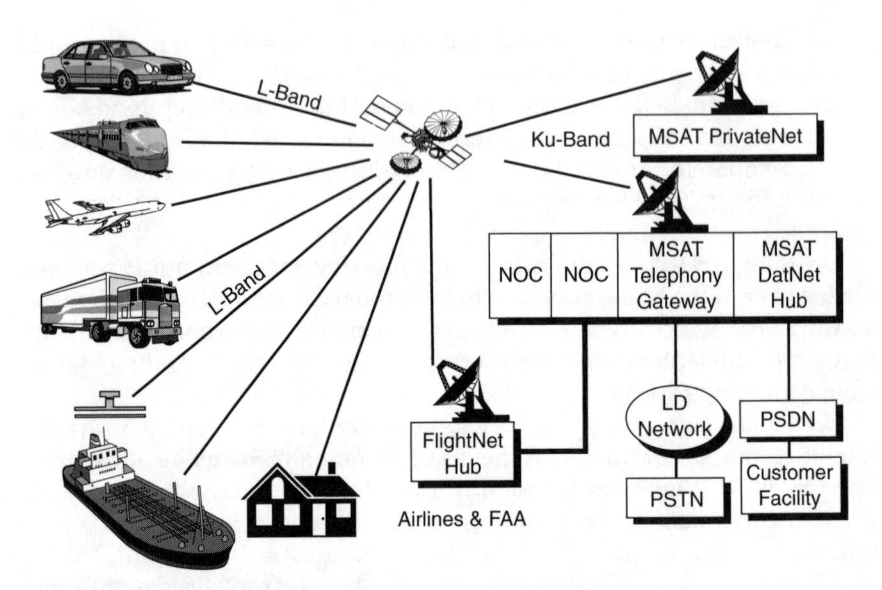

Figure 8.7 MSAT concept.

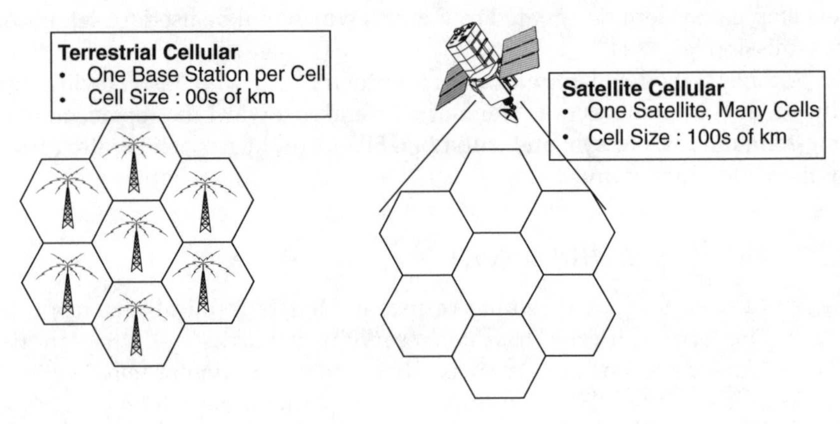

Figure 8.8 Comparison between terrestrial cellular and satellite cellular systems.

are compared in Figure 8.8.) First, the cells move very rapidly over the earth, and the mobile units, for all practical purposes, appear stationary — a kind of inverted cellular telephone system. Second, because of different designs, use of a hand-held unit is limited to the geographical coverage of a specific satellite constellation, and roaming of hand-held equipment among different satellite systems will not be allowed. MSS are extremely attractive since they can serve a significant sector of the cellular market — at least wherever cellular coverage is poor, thus augmenting the cellular coverage.

There are two types of constellation design approaches to satellite-based PCS. One approach is to provide coverage using three GEO satellites at approximately 36,000 km above the equator. The other approach involves using the LEO and MEO satellites at approximately 500 to 1500 km above the earth's surface. Thus, MSS are identified as either GEO or nongeostationary orbit (NGSO) satellites. The LEO and MEO satellites provide lower attenuation to the uplink and downlink signals in addition to lower signal delays because they operate at a lower altitude than the GEO satellites. Therefore, the NGSO satellites are emerging as major players in the world of wireless and personal communications.[20]

The main purpose of MSAT or MSS is to provide data and/or voice services to a fixed or portable personal terminal, close to the size of today's terrestrial cellular phones, by means of interconnection via satellite. LEO and MEO satellites have been proposed as an efficient way to communicate with these hand-held devices. The signals from the hand-held devices are retransmitted via a satellite to a gateway (a fixed earth station) which routes the signals through the public switched telephone network (PSTN) to its final destination or to another hand-held device.

Initial frequency bands were set aside in 1992 by the World Administrative Radio Conference (WARC). The L-band spectrum from 1610 to 1626.5 MHz was internationally allocated for MSS for earth-to-space (uplink) on a primary basis in all three ITU regions. WARC also allocated to MSS the band 1613.8 to 1626.5 MHz on a secondary basis and the spectrum in the S-band from 2483.5 to 2500 MHz on a primary basis for space-to-earth (downlink). Frequency allocation is accomplished on a regional basis. In the U.S., communications between the satellites and the fixed stations use Ku-band frequencies, while satellite-to-mobile links use the L-band frequencies. A reflector antenna with linear polarization can be used for the Ku-band links, while circular polarization can be used for the L-band satellite-to-mobile links.

8.5.1 Sample architectures

A sample of satellite systems designed for personal communications is now provided. These are the Iridium, Globalstar, and ICO systems.[21–24] All are global systems covering everywhere on earth. Each of these is characterized by two key elements: a constellation of non-geosynchronous satellites (LEO or MEO) arranged in multiple planes as shown in Figure 8.9, and a handheld terminal (handset) for accessing PCS, as shown in Figure 8.10.

8.5.1.1 Iridium

MSS reached a turning point in 1992 when Motorola introduced the concept of a LEO satellite system capable of directly serving handheld terminals. Motorola based its design of its big LEO constellation, known as Iridium, on the classical work of Adams and Ryder.[25] Iridium (www.iridium.com) is the first mobile satellite telephone network to offer voice and data services to and from hand-held telephones anywhere in the world. It uses a network of intersatellite switches for global coverage and GSM-type technology to link mobile units to the satellite network. As shown in Figure 8.11, the Iridium

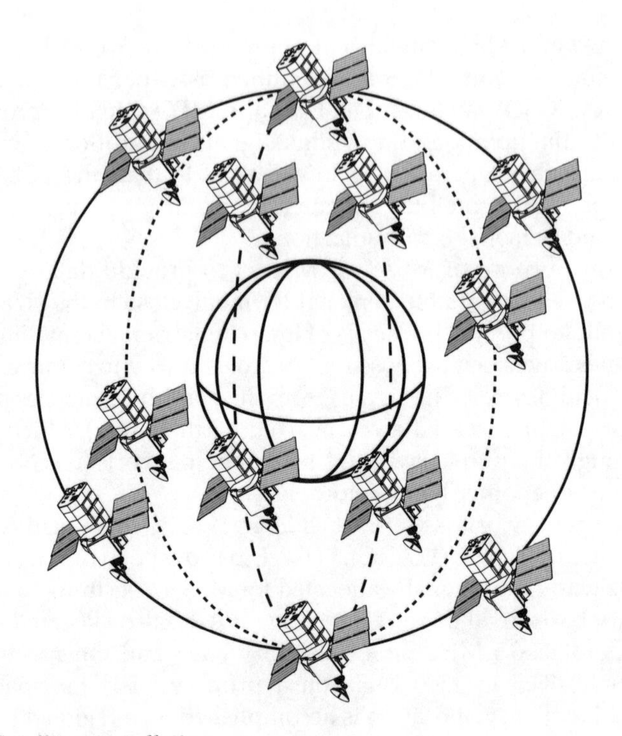

Figure 8.9 Satellite constellation.

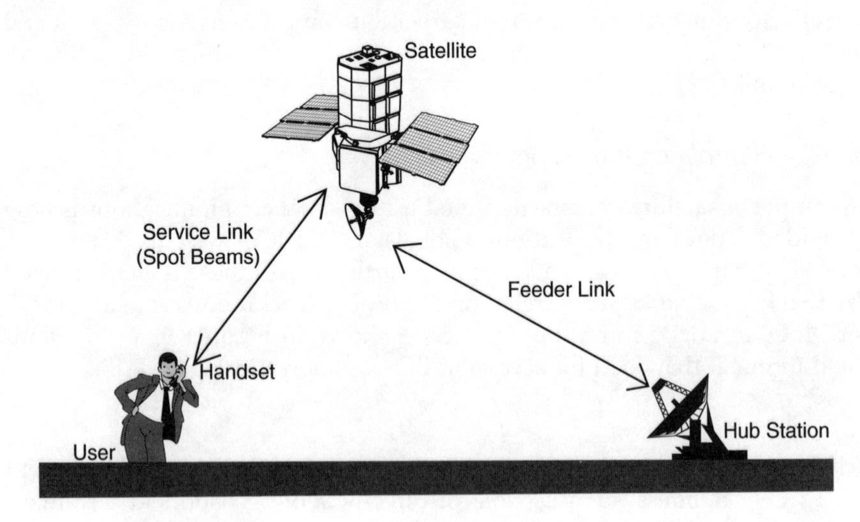

Figure 8.10 Ground segment.

system has three antenna arrays, each producing 16 cellular spot beams (cells, or rather megacells), and the beams are juxtaposed within the coverage area. The 48 spot beams travel very rapidly along the surface of the earth and the mobile units appear stationary.

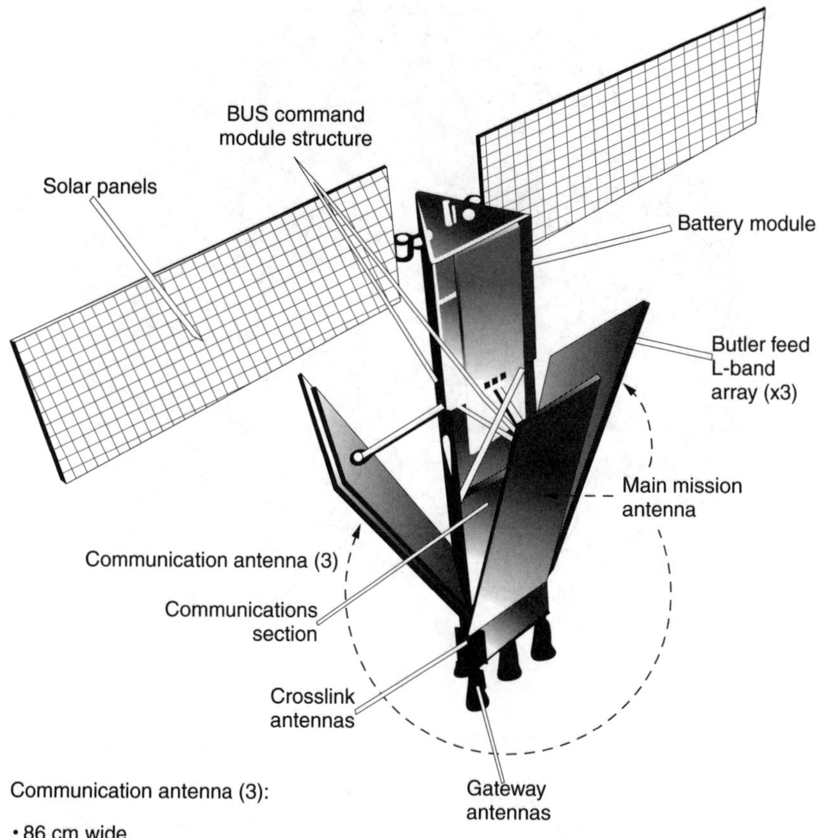

Figure 8.11 The Iridium satellite design.

Several modifications have been made to the original idea, including reducing the number of satellites from 77 to 66 by eliminating one orbital plane. (The name Iridium was based on the fact that the Iridium atom has 77 electrons.) Some of the key features of the current Iridium satellite constellation are listed below.[26–29]

- Number of (LEO) satellites: 66 (each weighing 700 kg or 1500 lb)
- Number of orbital planes: 6 (separated by 31.6° around the equator)
- Number of active satellites per plane: 11 (uniformly spaced, with one spare satellite per plane at 130 km lower in the orbital plane)
- Altitude of orbits: 780 km (or 421.5 nmi)
- Inclination: 86.4°

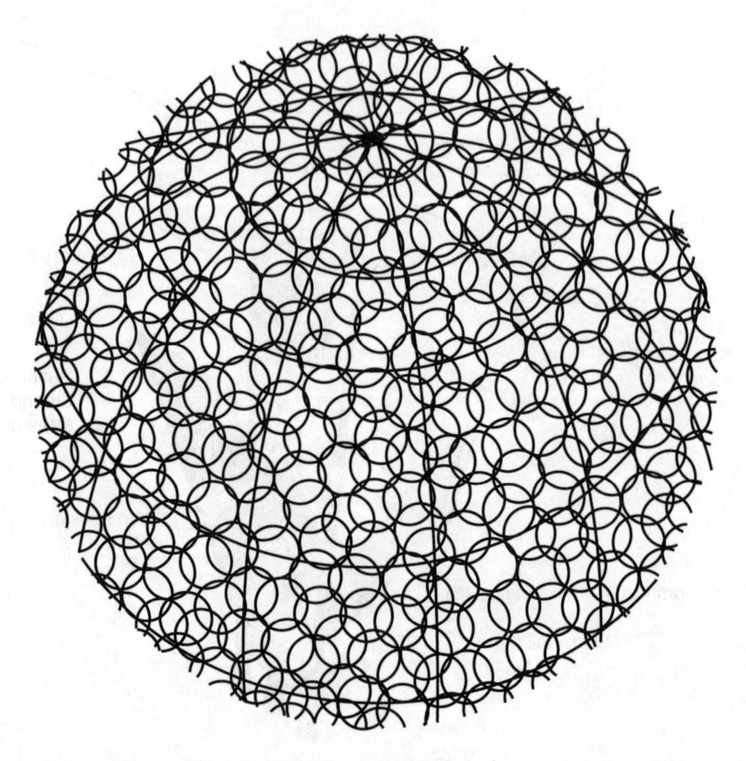

Figure 8.12 Iridium spot beams (cells) provide global coverage.

- Period of revolution: 100 min
- Design life: 8 years

Satellites in planes 1, 3, and 5 cross the equator in synchronization, while satellites in planes 2, 4, and 6 also cross in synchronization but out of phase with those in planes 1, 3, and 5. Collision avoidance is built into the orbital planning such that the closest distance between two satellites is 223 km. Each satellite covers a circular area about the size of the U.S. with a diameter of about 4400 km. The coverage area is divided into 48 cells. A typical spot beam coverage of the earth is shown in Figure 8.12. The satellites can project 48 spot beams using the L-band frequency assignments. Each of the spot beams is approximately 372 nmi in diameter.

The Iridium network uses FDMA/TDMA to produce efficient use of the spectrum; it provides voice at 4.8 kbps and data at 2.4 kbps. With FDMA, the available spectrum is subdivided into smaller bands assigned to individual users. Iridium extends this multiple access scheme further by using TDMA within each FDMA sub-band. Each user is assigned two time slots (one for sending and the other for receiving) within a repetitive time frame. During each time slot, the digital data are burst between the mobile handset and the satellite. The total spectrum of 5.15 MHz is divided into 120 FDMA channels with each satellite having about 1100 channels. Within each FDMA channel, there are four TDMA slots in each direction (uplink and downlink). Each TDMA slot is 8.29 ms in a 90-ms frame. The coded data burst rate with QPSK modulation and raised cosine filtering is 50 kbps. This design ensures the system will use less spectrum, keep channels closer without undue interference, and allow for acceptable levels of intermodulation.

The Iridium handsets are built by Motorola and Kyocera, a Japanese manufacturer of cellular telephones. The handsets are capable of both satellite access and terrestrial cellular roaming. Paging options are also available. The price of a typical handset is around $3000. The satellite service is about $3.00 per minute, which is about 25% more than the normal cellular roaming rate. The expected break-even market for Iridium is approximately 600,000 customers globally.

Outside the U.S., Iridium must obtain access rights in each country where service is provided. Iridium is negotiating to gain access to approximately 200 nations. Despite some problems expected of such a complex undertaking, Iridium has already been at work. Its 66 LEO satellites became fully commercial on November 1, 1998. However, on August 13, 1999, Iridium filed for bankruptcy and was later bought by Iridium Satellite LCC. Vendors competing with Iridium include Aries, Ellipso, and Globalstar.

8.5.1.2 Globalstar

Globalstar is a satellite-based cellular telephone system that allows users to talk from anyplace in the world. It will serve as an extension of terrestrial systems worldwide except for polar regions. The constellation is capable of serving up to 30 million subscribers.

Globalstar is being developed by the limited partnership of Loral Aerospace Corporation and Qualcomm with ten strategic partners. A functional overview of Globalstar is presented in Figure 8.13. Its key elements are:[30–32]

- *Space segment*: A constellation of 48 active LEO satellites located at an altitude of 1414 km and equally divided in eight planes (six satellites per plane). The satellite orbits are circular and are inclined at 52° with respect to the equator. Each satellite illuminates the earth at 1.6 GHz L-band and 2.5 GHz S-band with 16 fixed beams with service links, assignable over 13 FDM channels.
- *User segment*: Includes mobile and fixed users.
- *Ground segment*: Gateways (large ground station), ground operations control center (GOCC), satellite operations control center (SOCO), and Globalstar data network (GDN). The gateway enables communications to and from hand-held user terminals (UTs), relayed via satellite, with a public switched telephone network (PSTN). A gateway with a single radio channel transmits on a single frequency. The GOCC allocates capacity among gateways and collects operational control and billing information. The SOCC provides satellite maintenance and directs orbital maneuvers of the satellites.

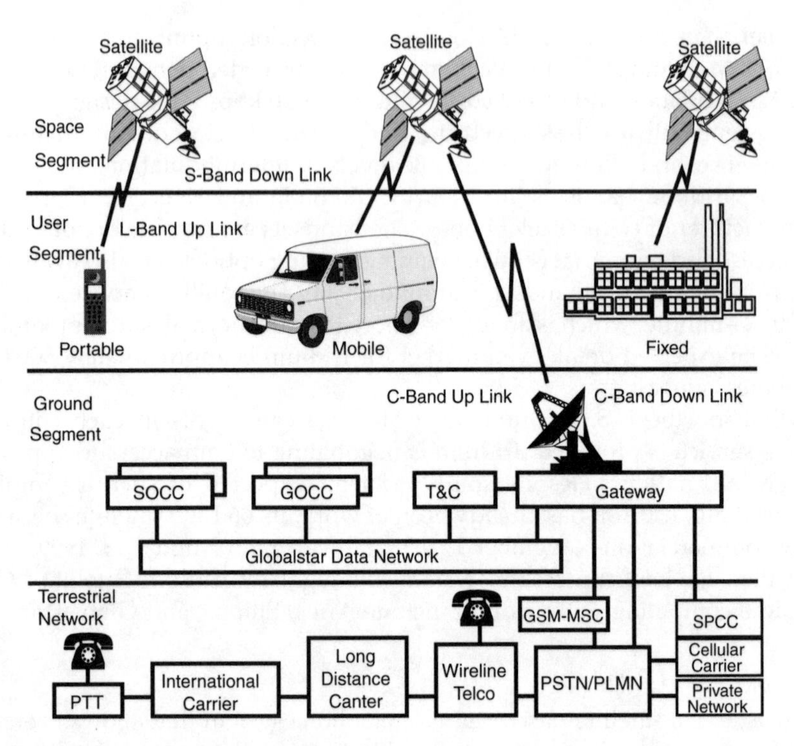

Figure 8.13 Globalstar system architecture.

The Globalstar satellites employ "bent pipe" transponders with the feeder link at C-band. Each satellite weighs about 704 lb and has a capacity of 2800 full-duplex circuits. It covers the earth with only 16 spot beams. Access to and from the satellite is at L- and S-bands, respectively, using CDMA in channels that are 1.25 MHz in bandwidth. The Qualcomm CDMA waveform employed in the user terminals and gateways spreads RF energy evenly over the allotted spectrum.

The Globalstar system offers the following services:

- One phone for both cellular and satellite calls
- Data and fax transmission
- Global roaming
- Short messaging service (SMS)
- Internet access

Since Globalstar plans to serve the military with commercial subscriptions, it employs signal encryption for protection from unauthorized calling parties. Unlike Iridium, which offers a global service, Globalstar's business plan calls for franchising its use to partners in different countries. For the latest information on the Globalstar system, visit the Web site at www.globalstar.com.

8.5.1.3 ICO

The ICO system is a satellite-based mobile communications system primarily to provide very cost effective voice, data, fax, and value-added services to handheld terminals. It will be used in a global satellite-based mobile communications system that will offer digital data and voice services as well as the satellite equivalent of third-generation wireless services such as wireless Internet and broadband services.

In Europe, Inmarsat in 1997 spun off a commercial organization known as Intermediate Circular Orbit (ICO) Global. The motivation was to use MEO rather than GEO as initially used by Inmarsat, thereby reducing the link propagation loss. The ICO system (originally called Inmarsat-P) was built by Hughes Space and Communications (now Boeing Satellite Systems). ICO constellation consists of:[33–35]

- 10 operational MEO satellites with 5 in each of the two inclined circular orbits at an altitude of 10,355 km
- One spare satellite in each plane, making 12 total launched
- Each satellite employs 163 spot beams
- Each satellite carries an integrated C- and S-band payload
- 12 satellite access nodes (SANs) located globally

The inclination of the orbits is 45° — making it the lowest of the systems described. Although this orbit reduces the coverage at high latitudes, it allows for the smallest number of satellites. The configuration is designed to provide coverage of the entire surface of the earth at all times and to maximize the path diversity of the system. ICO differs from Iridium and Globalstar in that it adopts the TDMA scheme for service links. Each satellite is designed to support at least 4500 telephone channels using TDMA.

An overview of the ICO system (www.ico.com) is shown in Figure 8.14. ICO is designed to provide the following services:

- Global paging
- Personal navigation
- Personal voice, data, and fax

The satellites will be linked on a terrestrial network known as the ICONET, which interconnects 12 ground stations, referred to as satellite access nodes (SANs). Each SAN consists of earth stations with multiple antennas for communicating with satellites, associated switching equipment, and a database to support mobility management. The orbital pattern of ICO is designed for significant coverage overlap to ensure that usually two (but sometimes three or four) satellites will be in view of a user and a SAN at any time. Each satellite will cover 25% of the earth's surface at any given time. The satellites will communicate with a terrestrial network through the ICONET, a high-bandwidth global IP network. The ICONET will include a system for managing

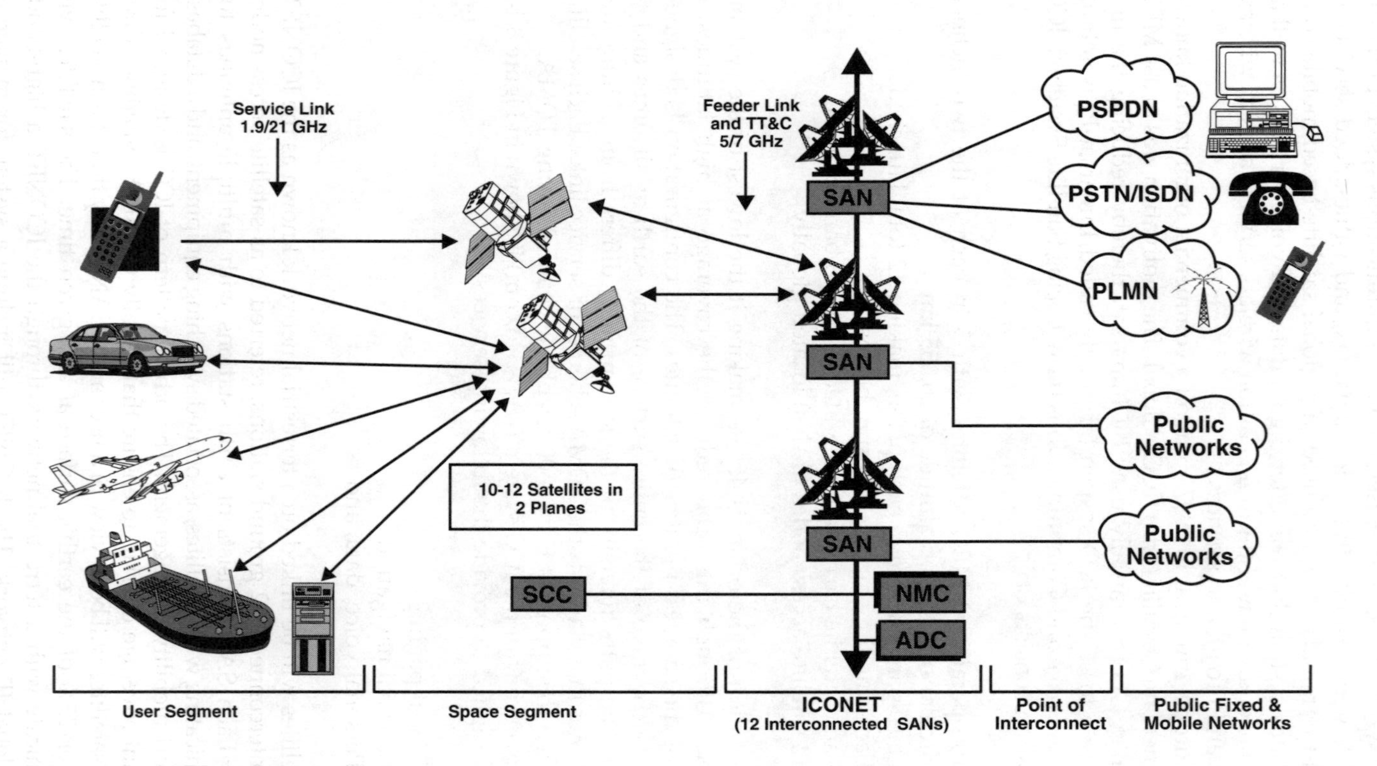

Figure 8.14 ICO system overview.

Table 8.5 Characteristics of Satellite PCS Systems

Parameter	Iridium	Globalstar	ICO
Company	Motorola	Loral/Qualcomm	ICO-Global
No. of satellites	66	48	10
No. of orbit planes	6	8	2
Altitude (km)	780	1414	10,355
Weight (lb)	1100	704	6050
Bandwidth (MHz)	5.15	11.35	30
Frequency Up/Down (GHz)	30/20	5.1/6.9	14/12
Spot beams/satellite	48	16	163
Carrier bit rate (kps)	50	2.4	36
Multiple access	TDMA/FDMA	CDMA/FDMA	TDMA/FDMA
Cost to build ($billion)	4.7	2.5	4.6
Service start date	1998	1999	2003

global user mobility based upon the existing digital cellular standard, GSM. The life span of ICO constellation is approximately 12 years. ICO is in the process of obtaining approvals from national and international regulatory authorities covering operation of the system and service provision.

The three constellations are compared in Table 8.5.

8.5.2 Applications

Applications of MSS include:

- *International travelers*: Satellites can provide services to people on the move. There is virtually no limit to the number of services that can be provided to the traveler, whether the travel be by land, sea, or air. The only possible constraint is that of a limited spectrum.
- *Global PCS*: MSS is designed to provide personal communication services (PCS) to those who need mobile communications in their own countries but who travel beyond the reach of terrestrial cellular systems.
- *Government agencies*: Government applications include law enforcement, fire, and public safety, among other services.
- *Broadband services*: Broadband satellite networks are the new generation of satellite networks in which Internet-based services will be provided to users regardless of their degree of geographical mobility. Still, voice, e-commerce, and low bit rate applications are among the services that will be provided.

As more customers subscribe to satellite mobile service (MSS), information will flow more freely, the world will grow smaller, and the global economy will be stimulated. The dawning age of global personal communication will bring the world community closer together as a single family.

Summary

Wireless communication is undergoing an explosive growth, and satellite-based delivery is a major player. With the introduction of satellite personal communications services in the near future, an important step will be taken toward the implementation of a global communication infrastructure.

Combining Kepler's laws with Newton's laws leads to a second-order differential equation. The solution to this equation yields the orbital parameters.

The go-anywhere-call-anywhere-from-anywhere concept will be achieved by means of satellite networks. Satellite-based service will use VSAT as the vehicle to deliver this global connectivity. VSATs are increasingly being used for communication from a single point to multipoint in broadcast applications and from multipoint to a single point in data collection systems.

FSS provides communications between two or more fixed locations. It involves one or more satellites as intermediate relay points, while the earth stations, located on the ground, serve stationary users.

With the advent of LEO/MEO constellations, global personal service communications (PCS) can be provided. They are ideal for providing personal communications to remote areas where terrestrial cellular service is not accessible. The LEO constellation is exemplified by the Iridium system, which is a commercial communication network comprised of a constellation of 66 LEO satellites. The system employs the L-band using FDMA/TDMA to provide global communications services through portable handsets. The system, however, is not without its technical and financial challenges. It competes with Globalstar and ICO.

Satellite networks already offer mobile telephone and paging service, but customers want global multimedia and broadband services. Integrating satellite networks with terrestrial ISDN and the Internet has been proposed.[36–39] The asymmetrical nature of Web traffic suggests a good match to VSAT systems since the VSAT return link capacity would be much smaller than the forward link capacity. More information about satellite communications systems can be obtained from References 40 through 42.

References

1. D. J. Marihart, Communications technology guidelines for EMS/SCADA systems, *IEEE Trans. Power Delivery,* vol. 16, no. 2, April 2001, 181–188.
2. D. Calcutt and L. Tetley, *Satellite Communications: Principles and Applications,* Edward Arnold, London, 1994, 3–15, 321–387.
3. B. R. Elbert, *The Satellite Communication Applications Handbook,* Artech House, Norwood, MA, 1997, 3–27, 257–320.
4. W. Pritchard, Geostationary versus non-geostationary orbits, *Space Commun.,* vol. 11, 1993, 205–215.
5. T. T. Ha, *Digital Satellite Communications,* McGraw-Hill, New York, 1990, 1–30, 615–633.

6. T. S. Tycz, Fixed satellite service frequency allocations and orbit assignment procedures for commercial satellite systems, *Proc. IEEE,* vol. 78, no. 7, July 1990, 1283–1288.
7. Y. Hu and V. O. K. Li, Satellite-based Internet: a tutorial, *IEEE Commun. Mag.,* March 2001, 154–162.
8. Propagation special issue, *Int. J. Satellite Commun.,* vol. 19, no. 3, May/June 2001.
9. D. C. Hogg and T. S. Chu, The role of rain in satellite communication, *Proc. IEEE,* vol. 63, 1975, 1308–1331.
10. M. Richharia, *Satellite Communication Systems,* McGraw-Hill, New York, 1995, 16–49.
11. T. Pratt, *Satellite Communications,* John Wiley & Sons, New York, 1986, 11–51.
12. W. W. Wu et al., Mobile satellite communications, *Proc. IEEE,* vol. 82, no. 9, Sept. 1994, 1431–1448.
13. B. R. Elbert, *Introduction to Satellite Communication,* Artech House, Norwood, MA, 1999, 390–395.
14. N. J. Muller, *Mobile Telecommunications Factbook,* McGraw-Hill, New York, 1998, 336–363.
15. M. H. Hadjitheodosiou et al., Next generation multiservice VSAT networks, *Electr. Comm. Eng. J.,* June 1997.
16. N. Abramson, Internet access using VSATs, *IEEE Commun. Mag.,* July 2000, 60–68.
17. N. Abramson, VSAT data networks, *Proc. IEEE,* vol. 78, no. 7, July 1990, 1267–1274.
18. J. C. Raison, Television via satellite: convergence of the broadcasting-satellite and fixed-satellite service — the European experience, *Space Commun.,* vol. 9, 1972, 129–141.
19. P. Wood, Mobile satellite services for travelers, *IEEE Commun. Mag.,* Nov. 1991, 32–35.
20. F. Abrishamkar, PCS global mobile services, *IEEE Commun. Mag.,* Sept. 1996, 132–136.
21. G. Comparetto and R. Ramirez, Trends in mobile satellite technology, *Computer,* Feb. 1997, 44–52.
22. J. V. Evans, Satellite systems for personal communications, *IEEE Antennas Propagation Mag.,* vol. 39, no. 3, June 1997, 7–20.
23. J. V. Evans, Satellite systems for personal communications, *Proc. IEEE,* vol. 39, no. 3, June 1997, 7–20.
24. Satellite communications — a continuing revolution, *IEEE Aerospace Electr. Sys. Mag.,* Oct. 2000, 95–107.
25. W. S. Adams and I. Rider, Circular polar constellations providing continuous single or multiple coverage above a specified latitude, *J. Astronaut. Sci.,* vol. 35, no. 2, April/June 1967.
26. B. Pattan, *Satellite-Based Cellular Communications,* McGraw-Hill, New York, 1998, 45–88.
27. P. Lemme et al., Iridium: Aeronautical satellite communications, *IEEE AES Syst. Mag.,* Nov. 1999, 11–16.
28. Y. C. Hubbel, A comparison of the Iridium and AMPS systems, *IEEE Network,* March/April 1997, 52–59.
29. R. J. Leopold and A. Miller, The Iridium communications system, *IEEE Potentials,* April 1993, 6–9.

30. E. Hirshfield, The Globalstar system: breakthroughs in efficiency in microwave and signal processing technology, *Space Commun.*, vol. 14, 1996, 69–82.
31. F. J. Dietrich et al., The Globalstar cellular satellite system, *IEEE Trans. Antennas Propagation*, vol. 46, no. 6, June 1998, 935–942.
32. R. Hendrickson, Globalstar for the military, *Proc. MILCOM*, vol. 3, 1998, 808–813.
33. P. Poskett, The ICO system for personal communications by satellite, *Proc. IEE Colloq.* (Digest), Part 1, 1998, 211–216.
34. L. Ghedia et al., Satellite PCN — the ICO system, *Int. J. Sat. Commun.*, vol. 17, 1999, 273–289.
35. M. Werner, Analysis of system parameters for LEO/ICO-satellite communication networks, *IEEE J. Selected Areas Commun.*, vol. 13, no. 2, Feb. 1995, 371–381.
36. T. Otsu et al., Satellite communication system integrated into terrestrial ISDN, *IEEE Trans. Aerosp. Electr. Sys.*, vol. 36, no. 4, Oct. 2000, 1047–1057.
37. C. Metz, TCP over satellite … the final frontier, *IEEE Internet Computer*, 76–80.
38. H. K. Choi, Interactive web service via satellite to the home, *IEEE Commun. Mag.*, March 2001, 182–190.
39. Y. Hu and V. O. K. Li, satellite-based internet: a tutorial, *IEEE Commun. Mag.*, March 2001, 154–162.
40. Special issue on satellite communications, *Proc. IEEE*, March 1977, vol. 65, no. 3.
41. Special issue on satellite communication networks, *Proc. IEEE*, Nov. 1984, vol. 72, no. 11.
42. Special issue on Global satellite communications technology and system, *Space Commun.*, vol. 16, 2000.

Problems

8.1 What advantages do satellite-based communications have over terrestrial communications?

8.2 What are the benefits of using GEO satellites for commercial purposes? Why are there objections to GEO satellites?

8.3 How do LEO satellites differ from GEO satellites? What are the advantages of the former?

8.4 State Kepler's laws of planetary motion.

8.5 A satellite has the following orbital parameters:

Semimajor axis: 42,168 km
Mean anomaly: 28.4°
Eccentricity: 0.00034
Calculate: (a) the maximum and minimum distances the satellite could be from the center of the earth during each revolution, (b) the orbital period in hours, (c) the mean orbital velocity.

8.6 The orbit of a satellite has a semimajor axis of 9000 km and an eccentricity of 0.125. Determine the values of (a) the latus rectum, (b) the minor axis, (c) the distance between foci.

8.7 A constellation is designed so that the earth central angle is 18.44°. How many satellites are in the constellation?

8.8 Compare and contrast the mesh and star topologies of a VSAT network.

8.9 What are the major limitations of VSAT networks? How do these limitations affect the applicability of VSAT networks to BISDN/ATM?

8.10 Name some applications of fixed satellite services.

8.11 Discuss briefly the need for MSS or MSAT systems and compare MSAT systems with the existing terrestrial cellular networks.

8.12 Explain the operation of Iridium.

8.13 Discuss the Globalstar system and its limitations.

8.14 How is ICO different from Iridium and Globalstar?

Bibliography

A. General

R. C. Dorf, *The Electrical Engineering Handbook*, CRC Press, Boca Raton, FL, 1997.
D. Minoli, *Enterprise Networking: Fractional T1 to Sonet, Frame Relay to BISDN*, Artech House, Boston, 1993.
M. N. O. Sadiku, *Wide and Metropolitan Area Networks*. Prentice-Hall, Upper Saddle River, NJ, in press.

B. Optical networks

G. P. Agrawal, *Fiber-Optic Communication Systems*, 2nd ed., John Wiley & Sons, New York, 1997.
American National Standard, Fiber Distributed Data Interface (FDDI) Token Ring Media Access Control (MAC), ANSI Standard X3.139, 1987.
American National Standard, Fiber Distributed Data Interface (FDDI) Physical Layer Protocol (PHY), ANSI Standard X3.148, 1988.
American National Standard, Fiber Distributed Data Interface (FDDI) Physical Layer Medium Dependent (PMD), ANSI Standard X3.166, 1990.
American National Standard, Fiber Distributed Data Interface (FDDI) Hybrid Ring Control (HRC), ANSI Standard X3.186, 1992.
American National Standard, Fiber Distributed Data Interface (FDDI) Station Management (SMT), ANSI Standard X3.229, 1994.
R. Bartnikas and K. D. Srivastava (eds.), *Power and Communication Cables: Theory and Applications*, IEEE Press, Piscataway, NJ, 2000.
T. Beninger, *SONET Basics*, Telephony, Chicago, 1991.
A. F. Benner, *Fiber Channel*, McGraw-Hill, New York, 1996.
A. H. Cherin, *An Introduction to Optical Fibers*, McGraw-Hill, New York, 1983.
B. Chomycz, *Fiber Optic Installations*, McGraw-Hill, New York, 1996.
M. C. Chow, Understanding SONET/SDH: Standards and Applications, Andan Publisher, Holmdel, NJ, 1995.
R. P. Davidson and N. J. Muller, *The Guide to SONET: Planning, Installation, and Maintaining Broadband Networks*, Telecom Library, New York, 1991.
H. J. R. Dutton, *Understanding Optical Communications*, Prentice-Hall, Upper Saddle River, NJ, 1998.
T. Edwards, *Fiber-Optic Systems*, John Wiley & Sons, Chichester, 1989.
G. Einarsson, *Principles of Lightwave Communications*, John Wiley & Sons, Chichester, 1996.
J. Enck and M. Beckman, *LAN to WAN Interconnection*, McGraw-Hill, New York, 1995.
C. Flanigan and S. Sarker, *WDM: Global Strategies for Next Generation Networks*, Ovum, London, 1998.
W. J. Goralski, *SONET: A Guide to Synchronous Optical Network*, McGraw-Hill, New York, 1997.
P. E. Green, *Fiber Optic Networks*, Prentice-Hall, Englewood Cliffs, NJ, 1993.

R. J. Hoss, *Fiber Optic Communications: Design Handbook*, Prentice-Hall, Englewood Cliffs, NJ, 1990.

R. Jain, *FDDI Handbook: High-Speed Networking Using Fiber and Other Media*, Addison-Wesley, Reading, MA, 1994.

W. B. Jones, *Introduction to Optical Fiber Communication Systems*, Rinehart and Winston, New York, 1987.

L. Kazovsky, S. Benedetto, and A. Willner, *Optical Fiber Communication Systems*, Artech House, Boston, 1996.

G. Keiser, *Optical Fiber Communications*, 3rd ed., McGraw-Hill, New York, 2000.

H. B. Killen, *Fiber Optic Communications*. Prentice-Hall, Englewood Cliffs, NJ, 1991.

J. A. Kuecken, *Fiberoptics: A Revolution in Communication*, 2nd ed., Tab Books, Blue Ridge Summit, PA, 1987.

S. L. W. Meardon, *The Elements of Fiber Optics*, Prentice-Hall, Englewood Cliffs, NJ, 1993.

J. E. Midwinter and Y. L. Guo, *Optoelectronics and Lightwave Technology*, John Wiley & Sons, Chichester, U.K., 1992.

S. E. Miller and A. G. Chynoweth (Eds.), *Optical Fiber Communications*, Academic Press, Orlando, FL, 1979.

B. Mukherjee, *Optical Communication Networks*, McGraw-Hill, New York, 1997.

K. Nosu, *Optical FDM Network Technologies*, Artech House, Boston, 1997.

J. C. Palais, *Fiber Optic Communications*, 3rd ed., Prentice-Hall, Englewood Cliffs, NJ, 1992.

D. Papannareddy, *Introduction to Lightwave Communication Systems*, Artech House, Boston, 1997.

J. P. Powers, *An Introduction to Fiber Optic Systems*, Irwin, Boston, 1993.

R. Ramaswami and K. N. Sivarajan, *Optical Networks: A Practical Perspective*, Morgan Kaufmann Publishers, San Francisco, 1998.

R. J. Ross and E. A. Lacy, *Fiber Optics*, 2nd ed., Prentice-Hall, Englewood Cliffs, NJ, 1993.

R. Sabella and P. Lugli, *High Speed Optical Communications*, Kluwer Academic Publishers, Dordrecht, 1999.

R. G. Seippel, *Optoelectronics*, Reston Publishing, Reston, VA, 1981.

A. Shah and G. Ramakrishnan, *FDDI: A High Speed Network*, Prentice-Hall, Englewood Cliffs, NJ, 1993.

M. J. N. Sibley, *Optical Communications*, McGraw-Hill, New York, 1990.

D. M. Spirit and M. J. O'Mahony, *High Capacity Optical Transmission Explained*, John Wiley & Sons, Chichester, 1995.

G. R. Stephens and J. V. Dedek, *Fiber Channel: The Basics*, ANCOT Corporation, Menlo Park, CA, 1997.

T. E. Stern and K. Bala, *Multiwavelength Optical Networks*, Addison-Wesley, Reading, MA, 1999.

N. Thorsen, *Fiber Optics and the Telecommunications Explosion*, Prentice-Hall, Upper Saddle River, NJ, 1998.

M. Young, *Optics and Lasers*, 3rd ed., Springer-Verlag, Berlin, 1986.

C. Wireless networks

J. Agosta and Travis Russel, *CDPD: Cellular Digital Packet Data Standards and Technology*, McGraw-Hill, New York, 1997.

B. Bates, *Wireless Networked Communications*, McGraw-Hill, New York, 1993.

D. Calcutt and L. Tetley, *Satellite Communications: Principles and Applications*, Edward Arnold, London, 1994.

P. T. Davis and C. R. McGuffin, *Wireless Local Area Networks*, McGraw-Hill, New York, 1995.

R. A. Dayem, *Mobile Data and Wireless LAN Technologies*, Prentice-Hall, Upper Saddle River, NJ, 1997.

B. R. Elbert, *Introduction to Satellite Communication*, Artech House, Norwood, MA, 1999.

B. R. Elbert, *The Satellite Communication Applications Handbook*, Artech House, Norwood, MA, 1997.

S. Faruque, *Cellular Mobile Systems Engineering*, Artech House, Norwood, MA, 1996.

K. Feher, *Wireless Digital Communications*, Prentice-Hall, Upper Saddle River, NJ, 1995.

V. K. Garg, K. Smolik, and J. E. Wilkes, *Applications of CDMA in Wireless/Personal Communications*, Prentice-Hall, Upper Saddle River, NJ, 1997.

V. K. Garg and J. E. Wilkes, *Wireless and Personal Communications Systems*, Prentice-Hall, Upper Saddle River, NJ, 1996.

J. B. Gibson (Ed.), *The Mobile Communications Handbook*, 2nd ed., CRC Press, Boca Raton, FL, 1999.

M. Golio (Ed.), *The RF and Microwave Handbook*, CRC Press, Boca Raton, FL, 2001.

D. J. Goodman, *Wireless Personal Communications Systems*, Prentice-Hall, Upper Saddle River, NJ, 1997.

T. T. Ha, *Digital Satellite Communications*, McGraw-Hill, New York, 1990.

W. C. Y. Lee, *Mobile Communications Engineering*, 2nd ed., McGraw Hill, New York, 1997.

N. J. Muller, *Mobile Telecommunications Factbook*, McGraw-Hill, New York, 1998.

N. J. Muller, *Wireless Data Networking*, Artech House, Boston, 1995.

M. Nemzow, *Implementing Wireless Networks*, McGraw-Hill, New York, 1995.

K. Park, *Personal and Wireless Communications: Digital Technology and Standards*, Kluwer Academic Publishers, Boston, 1996.

B. Pattan, *Satellite-Based Cellular Commuications*, McGraw-Hill, New York, 1998.

D. M. Pozar, *Microwave and RF Design of Wireless Systems*, John Wiley & Sons, New York, 2001.

T. Pratt, *Satellite Communications*, John Wiley & Sons, New York, 1986.

F. J. Ricci, *Personal Communications Systems Applications*, Prentice-Hall, Upper Saddle River, NJ, 1997.

F. S. Rappaport, *Wireless Communications: Principles and Practice*, Prentice-Hall, Upper Saddle River, NJ, 1996.

T. S. Rappaport (Ed.), *Cellular Radio and Personal Communications*, IEEE, Piscataway, NJ, 1995.

F. J. Ricci, *Personal Communications Systems Applications*, Prentice-Hall, Upper Saddle River, NJ, 1997.

M. Richharia, *Satellite Communication Systems*, McGraw-Hill, New York, 1995.

C. Smith and C. Gervelis, *Cellular System: Design and Optimization*, McGraw-Hill, New York, 1996.

M. Sreetharan and R. Kumar, *Cellular Digital Packet Data*, Artech House, Norwood, MA, 1996.

R. Steele and L. Hanzo, *Mobile Radio Communications*, 2nd ed., John Wiley & Sons, Chichester, 1999.

W. Webb, *Introduction to Wireless Local Loop: Broadband and Narrowband Systems*, 2nd ed., Artech House, Boston, 2000.

Glossary and acronymns

8B/10B: A type of encoding/decoding of bytes to reduce errors in transmission.

Absorption: The soaking up of light, heat, sound, and other forms of energy by a substance.

Absorption loss: Attenuation of optical signal caused by the atoms that constitute the fiber.

Acceptance angle: The maximum angle above which light will not enter the fiber.

Adjacent channel: A carrier frequency ±30 kHz from the current carrier frequency in North American cellular systems or ±200 kHz from the current carrier for GSM and related systems.

ADM: Add/drop multiplexer; an electronic device that provides opto-electric/electro-optic conversion allowing adding, dropping, or multiplexing of signals.

AMPS: Advanced mobile phone service; an analog cellular standard operating in the frequency range of 800 MHz with a bandwidth of 30 kHz; it is used in North America and other parts of the world.

ANSI: American National Standards Institute.

AON: All-optical network; a wavelength-routed optical network in which transmission functionality such as multiplexing, routing, or switching is implemented through all-optical processing rather than through electronic processing.

APD: Avalanche photodiode; a type of photodiode that takes advantage of the avalanche multiplication of photocurrent.

Apogee: The point in the satellite orbit that is farthest from the center of the earth.

Application layer: This is layer 7 (highest layer) in the OSI protocol hierarchy. It provides an interface between user programs and data communication.

Architecture: The general design of hardware or software, including how they fit together.

ARQ: Automatic repeat request; an error correction technique whereby blocks that are determined to be in error are re-sent.

ASII: American Standard Code for Information Interchange.

Asynchronous transmission: A transmission method with no clocking signal.

ATIS: Alliance for Telecommunication Industry Solutions; the U.S. body responsible for the standardization of GSM-based solutions for use in the U.S.

ATM: Asynchronous transfer mode; packetized multiplexing scheme that allows services of different bit rates to be efficiently carried on an optical fiber. It is the broadband ISDN mode similar to asynchronous time division multiplexing.

Attenuation: Reduction in magnitude of current, voltage, or power of a signal.

Backbone: A network that links two or more other local networks.

Band gap: Also known as energy gap; the amount of energy required for an electron to jump from the valence band to the conduction band.

Bandwidth: The range of frequencies that are passed by a communications channel without significant attenuation.

Base station: A low transceiver equipment located in each cell in a cellular service area.

Baseband: A method of transmitting signals without modulation.

BER: Bit error rate; the probability of receiving bits in error in the course of transmission.

BISDN: Broadband integrated services digital network; standards being developed for ISDN to handle services, such as video, requiring high bandwidth.

Bit: Abbreviation for binary digit.

Bit rate: Number of bits per second transmitted or received; it is the minimum speed at which information must be sent for satisfactory reception.

Bps: Bits per second.

Broadband: A term used to describe a communication service that delivers communication channels with a bandwidth of 1.5 Mbps or higher.

Broadcast: Simultaneous transmission of signal to a number of nodes.

Broadcast-and-select network: A network in which the information transmitted by a node passes through all other nodes and each node selects which part of the information it wants to receive.

CATV: Community antenna television; a communications service that provides TV signals to a large community through cable.

CCITT: Consultative Committee on International Telegraphy and Telephony; an international body that develops telecommunications standards; now known as ITU-T.

CDMA: Code-division multiple access; multiple channels on the same wavelength but separated by the way they encode data.

CDPD: Cellular digital packet data; an open protocol for transmitting data over the existing AMPS cellular infrastructure at 19.2 k baud.

Cell: The basic geographic unit of a cellular system.

Cell splitting: Dividing one cell into two or more cells to provide additional capacity within the original cell's region of coverage.

Cellular: A system of reusing bandwidth by dividing a region into small cells, each with a stationary radio antenna.

CEPT: European Conference of Postal and Telecommunications Administration; the European spectrum manager.

Channel: A communication path between two or more points of termination.

Circuit switching: A switching mechanism that establishes a dedicated path between users that is held for the duration of the communication. Most telephone connections are circuit-switched.

Cladding: A material with low refractive index that surrounds the core and provides optical insulation and protection of the core.

Coaxial cable: A transmission medium consisting of an insulated core surrounded by a braided shield.

Congestion: A slow-down in a network due to a bottleneck or excessive traffic.

Connection-oriented transmission: Data transmission involving setting up a connection before transmission and disconnection after all data has been transmitted. It is analogous to a telephone call.

Connectionless transmission: Data transmission without setting up a connection. It is analogous to posting a letter.

Core: The central dielectric portion of an optical fiber through which the optical wave propagates.

Coupler: A device that is designed to split optical power into other fibers.

CPA: Combined paging and access (field); a bit field in the forward overhead message which informs the mobile unit if the system has combined paging and access channels.

CRC: Cyclic redundancy check; an error detection scheme that applies a special algorithm to a series of bytes and calculates a numeric value that is appended to the data.

Critical angle: The least angle of incidence at which total internal reflection takes place.

Crosstalk: The undesirable effect of a transmission on one channel interfering with the transmission on another channel.

CSMA: Carrier sense multiple access; a medium mechanism used in LANs.

CSMA/CA: Carrier sense multiple access with collision avoidance; an access protocol used by wireless LAN.

CSMA/CD: Carrier sense multiple access with collision detection; it is CSMA with sensing during transmission to detect collisions.

CT1: Cordless telephone (first generation); an analog cordless telephone standard operating at a frequency range of 46–49 MHz, with a bandwidth of 25 kHz.

CT2: Cordless telephone (second generation); a system based on ETSI standard that describes a residential cordless telephone that can be used commercially.

Data link layer: Layer 2 of the OSI reference model.

DCS: Digital communications services; a part of the GSM digital cellular standard that operates in the higher bands around 1800 GHz.

De facto standard: A default standard; a standard by usage rather than official decree.

De jure standard: An official standard.

Dead spot: An area within the service area where the radio signal strength is significantly reduced.

DECT: Digital enhanced cordless telephone; a cordless telephone standard developed in Europe but likely to receive worldwide acceptance for wireless PBXs, residential, and public cordless applications.

Delay: Transfer delay is the time between the arrival of a packet to a node interface and its complete delivery at the destination node.

Demultiplexer: A device that separates signals that have been multiplexed together.

Detector: A device that converts optical energy to electrical energy.

Digital color code (DCC) [field]: A bit field whose value corresponds to one of the three SAT codes — the one assigned to the local cell site.

Dispersion: A broadening of input pulses along the length of the fiber.

Distortion: A change in the shape of the signal.

Downstream: Node or station immediately next to the reference one in the transmission path.

DQDB: Distributed queue dual bus; the IEEE 802.6 standard MAN.

DSSS: Direct sequence spread spectrum; a spread-spectrum modulation scheme that divides the available bandwidth into three or four subchannels.

DTM: Dynamic synchronous transfer mode.

DTMF: Dual tone multi-frequency; a tone-dialing system based on outputting two nonharmonic related frequencies simultaneously to identify the number dialed.

DWDM: Dense wavelength division multiplexing; a term used to describe recently developed WDM systems with very high channel capacities.

EDFA: Erbium-doped fiber amplifier; an amplifier that can amplify signals in the wavelength range of 1540 nm to 1570 nm, which is one of the low-loss operating regions of the optical fiber.

EMI: Electromagnetic interference.

Ethernet: A bus-based LAN using CSMA/CD medium access protocol.

Fairness: Equitable treatment of all users, especially in terms of access to a network.

FCC: Federal Communications Commission; a regulatory authority for telecommunications in the U.S.

FDDI: Fiber-distributed data interface; a shared-medium token-passing LAN/MAN technology based on fiber optic links operating at 10 Mbps.

FDM: Frequency division multiplexing; a multiplexing scheme that assigns different frequencies to different signals and sends them on a single channel.

FEC: Forwarding equivalent class.

FES: Fixed-end system; an end system that provides application-level services, but it is not mobile.

FHSS: Frequency hopping spread spectrum; a spread-spectrum scheme that divides the available bandwidth into a large number of small subchannels.

Fiber: An optical transmission waveguide consisting of a core and a cladding.

Flow control: A means by which a receiving entity limits the amount of data sent by the transmitting entity.

Frame: The data unit of the MAC protocol; consists of the LLC data and the MAC header and trailer information.

Frequency hopping: A method of transmission where the channels are visited in a predefined order specified by a hopping sequence.

Frequency reuse: The ability to reuse channels on the same frequency without causing interference.

FSS: Fixed satellite service.

FTTB: Fiber to the business; the application of optical fiber to carry telecommunication signals to business premises.

FTTC: Fiber to the curb/customer; the distribution of communication services by providing optical fiber links to a central point in each neighborhood and continuing to the homes by either coax or twisted pair.

FTTH: Fiber to the home; the application of optical fiber to carry telecommunication signals directly to residential homes.

Full duplex: The ability to transfer information in both directions of the link simultaneously.

FVC: Forward analog voice channel; the analog voice or traffic channel that is from the base station to the mobile telephone.

GEO: Geosynchronous orbit; an orbit taken by satellite where the satellite's orbit velocity matches the rotation of the earth causing the satellite to remain stationary relative to a position on the earth's surface.

Graded-index fiber: An optical fiber whose refractive index is a function of the radial distance from the fiber's axis.

Ground station: A collection of communications equipment for receiving or transmitting from or to satellite.

GSM: Global system for mobile communications; the most successful digital cellular system in the world.

Half duplex: The ability to transfer information in both directions of the link but not at the same time.

Hand-off: Also called Handover in Europe; the transfer of responsibility for a call from one cell site to the next.

Header: Control information that precedes user data.

HDTV: High definition television; a new standard for television transmission consisting of 1080 lines by 1920 square pixels in a 16×9 digital format.

HIPPI: High performance parallel interface; an ANSI standard for high-speed transfer of information in a dual-simplex manner over a short parallel bus.

Hop: The passage of a packet from one switch, bridge, router, or gateway.

Hopping sequence: The preset order in which frequency-hopping RF transmissions are distributed over the 82 channels of the assigned ISM band.

Hot spots: Regions in a cellular service area that have much higher traffic than the average.

ICO: Intermediate circular orbit.

IEEE: Institute of Electrical and Electronics Engineers.

IEEE 802: The committee assigned by IEEE to provide standards for LANs:
IEEE 802.1 — standard for LAN/MAN bridging and management
IEEE 802.2 — standard for logical link control protocol.
IEEE 802.3 — standard for CSMA/CD protocol.
IEEE 802.4 — standard for token bus MAC protocol.
IEEE 802.5 — standard for token ring MAC protocol.
IEEE 802.6 — standard for metropolitan area networks
IEEE 802.7 — standard for broadband LAN
IEEE 802.8 — standard for fiber optics
IEEE 802.9 — standard for integrated services
IEEE 802.10 — standard for LAN/MAN security
IEEE 802.11 — standard for wireless LAN
IEEE 802.12 — standard for demand priority access method
IEEE 802.14 — standard for cable TV
IEEE 802.15 — standard for wireless personal area network (WPAN)
IEEE 802.16 — standard for broadband wireless access

Index of refraction: The ratio of the speed of light in a vacuum to the speed of light in a material.

Infrared: Light wavelengths extending from 770 nm on.

ILD: Injection laser diode; a semiconductor device that is capable of emitting coherent or stimulated radiation under specified conditions.

IP: Internet protocol; a set of protocols developed by the U.S. Department of Defense to communicate between computers across networks.

IPI: Intelligent peripheral interface; an ANSI standard for controlling peripheral devices by a host computer.

ISDN: Integrated services digital network; an end-to-end digital network that supports a wide range of services accessed by a limited set of multipurpose user-network interfaces.

ISM-band: Industrial, scientific, and medical band; the portion of the frequency spectrum allocated to industrial, scientific, and medical wireless applications — 902–928 MHz, 2400–2483 MHz, and 5.7–5.9 GHz.

ISO: International Standards Organization; an international body responsible for setting standards.

ITU-T: International Telecommunications Union Telecommunication sector; the standardization body for telecommunication networks and services.

LAN: Local area network; a computer network that interconnects telecommunications devices in a small geographical area such as a building or campus.

Laser: Light amplification by simulated emission of radiation; a semiconductor device used as a coherent source of light in an optical fiber system.

LATA: Local access and transport area; the geographical domain of the local exchange carrier.

Layer: A conceptual region that embodies some functions between an upper logical boundary and a lower logical boundary.

LDP: Label distribution protocol.

Lease line: Also called private line; a dedicated common carrier circuit providing a point-to-point or multipoint network connection.

LEC: Local exchange carrier; a telephone service provider that furnishes local telephone service to end users.

LED: Light emitting diode; a pn junction device that radiates incoherent light when biased in the forward direction.

LEO: Low earth orbit; a satellite orbit between 500 km and 1000 km above the surface of the earth, used by new generation of personal communication satellites.

LER: Label edge router.

LLC: Logical link control; the upper part of the data link layer.

LOS: Line of sight; the connection between communication devices (such as microwave, laser, and infrared systems) in which there are no obstructions on the direct path between transmitter and receiver.

LSR: Label switch router.

MAC: Medium access control; a lower part of the Data Link layer that ensures that only one node has access to the medium at a time.

MAN: Metropolitan area network; a computer network that links a number of telecommunications devices within a large geographical area such as a city.

Mbps: Megabits per second; 2^{20} or 1,048,576, roughly 1 million bits per second.

MDBS: Mobile data base station; the CDPD channel stream controller.

MDIS: Mobile data intermediate system; provides serving functions to mobile-end systems and authentication or forwarding services for client MESs.

MEO: Medium earth orbit.

MES: Mobile-end system; the CDPD mobile device which must register and be granted permission to use the network service after authentication has been verified.

Microwave: An electromagnetic wave in the frequency range of 3 to 30 GHz.

Mode: An electromagnetic field pattern or propagation of light energy allowable in the core of the fiber.

Modem: A modulator/demodulator that converts a digital bit stream into an analog signal (modulation) and vice versa (demodulation).

MONET: Multiwavelength optical network; a prototype WDM network funded in part by the Defense Advanced Research Projects Agency (DARPA) and operated in the U.S. by a consortium of vendors including AT&T, Bell Atlantic, Telcordia, BellSouth, SBC, and Lucent.

MPLS: Multiprotocol label switching; an extension of the existing IP architecture.

MSA: Metropolitan statistical area; an area designated by the FCC for service to be provided for by a cellular carriers.

MSS: Mobile satellite service.

MTSO: Mobile telephone switching office; the central coordinating element which is made up of signal processors, memories, switching networks, truck circuits, and ancillary services.

Multicasting: The ability to transmit from a single source to multiple destinations.

Multimode fiber: An optical fiber that supports propagation of more than one mode of light signal transmission; generally used for short-distance transmission.

Multipath: A condition where a signal from one source is received by several (a direct and many reflected) paths.

Multiplexer: A device that allows several channels to be carried over a single physical link or fiber.

Multiplexing: A process of combining separate communication channels into one composite channel.

Network: An arrangement of nodes and connecting branches.

Network element: Any component of a communications network.

Network layer: Layer 3 of the OSI reference model.

NIC: Network interface card; hardware installed in network devices that enables them to communicate on a network.

NMT: Nordic mobile telephone; an analog mobile phone standard used widely in Scandinavia at 450 and 900 MHz.

Node: A station or data communication equipment at a geographical location in a telecommunication network.

Numerical aperture: A number that defines the light-gathering capability of a fiber; a characteristic of an optical fiber that is specified in term of its acceptance angle.

OADM: Optical add-and-drop multiplexer; it can extract or insert one or more wavelength channels from a stream of channels in a WDM link without affecting other channels.

OC-*n*: OC indicates optical carrier and *n* represents multiples of 51.48 Mbps; a fundamental unit used in SONET hierarchy.

Optical fiber: Communication medium used for fiber optics.

Optical layer: A layer, also known as photonic layer, in which network functionality is carried out through entirely optical means.

OSI: Open Systems Interconnection; a model architecture and protocol hierarchy by ISO.

OTDM: Optical time division multiplexing; optical equivalent to electronic TDM.

PABX: Public access branch exchange; a switch in a fixed network.

Packet switching: A switching mechanism that involves breaking a message into packets which are routed independently of one another to their destination in a store-and-forward manner over multiple virtual circuits.

Paging: The process in which the base station sends a message over the control channel informing the mobile that a mobile destination is coming.

Payload: The actual information contents carried within a cell, frame, or packet; the portion of the data field in a cell, frame or packet besides optional headers.

PBX: Private branch exchange; a small switch within an office providing service to the users in the office, allowing interoffice calls, and providing a point of interconnection with the PSTN.

PCN: Personal communications network; the infrastructure or land network part of a PCS system.

PCS: Personal communication services; the next generation of wireless services that will provide subscribers with portable two-way voice, fax, e-mail, data, and image capabilities regardless of location.

PDH: Pleisiochronous digital hierarchy; an asynchronous multiplexing scheme from 1.5 Mbps to 565 Mbps.

Perigee: The point in a satellite orbit that is closest to the center of the earth.

Photodetector: A device that receives light energy and transforms it into electrical energy.

PHS: Personal handyphone system; a short-range TDMA/TDD radio-telephone system developed in Japan to be the 1.9 GHz residential cordless/wireless PBX system.

PHY: Physical layer.

Physical Layer: Layer 1 of the OSI reference model.

Picocell: Cell site smaller than the microcell.

PIN diode: A type of photodiode used in a receiver to convert optical signals into electrical signals.

Presentation layer: Layer 6 of the OSI reference model.

Propagation time: Time interval between signal transmission at the sending node and its reception at the receiving node.

Protocol: A formal set of rules for exchange of information between nodes.

PSTN: Public switched telephone network; a fixed telephone network such as that operated by the PTO.

PTO: Post and telecommunication organization; the company that operates the main PSTN within a country.

QPSK: Quadrature phase shift keying; an efficient modulation scheme which breaks the information path into two parts: the "in phase" and the "quadrature phase" components.

Quantum efficiency: The efficiency of the conversion of incoming photons to electron-hole pairs.

Random access: An uncontrolled access to transmission medium in which stations transmit when ready and later resolve any conflict that may arise.

Refractive index: See **Index of refraction.**

Regenerator/repeater: A device that retransmits a degraded digital signal for continued transmission.

Repeater: A physical layer device for amplifying or regenerating signals.

Responsivity: A parameter used to describe the sensitivity of a photodetector; it is the ratio of the output current to the input power.

Reuse of frequencies: The assigning of frequencies to cell sites so that no adjoining cell sites use the same ones; cell sites out of range of one another can use or reuse the same frequencies.

Roaming: Being able to access a network regardless of location and move freely while maintaining an active link through a wireless connection to a network; roaming usually requires a hand-off when a node (or user) moves from one cell to another.

Router: A network layer device that routes data between LANs.

Routing: Technique for transmitting a message from source to destination.

Scalability/modularity: The ability to scale or grow a network to a larger size in small increments (modules).

SCSI: Small computer system interface; an ANSI standard for controlling peripheral devices by one or more host computers.

SDH: Synchronous digital hierarchy; a set of ITU-T standards for synchronous digital transmission over optical fiber. It is the world-wide equivalent of SONET standard used in North America.

Session layer: Layer 5 of the OSI reference model.

Signal-to-noise ratio: A term used to describe the quality of a transmission system and defined as the ratio of the power of the signal to the power of the background noise.

Signaling: The procedure for creating and clearing connections.

SMDS: Switched multimegabit data service; a high-speed public packet data service used in MAN applications.

SMF: Single-mode fiber; an optical fiber that supports propagation of only a single mode of light signal transmission; dominant in long-distance transmission.

SONET: Synchronous optical network; an ANSI standard defined for synchronous digital transmission over optical fiber or a North American version of SDH.

Spread spectrum: A technique used in reducing and avoiding interference by taking advantage of statistical means to send a signal between two points.

SS7: Signaling system #7; an international standard networking signaling protocol which allows common channel signaling between telephone network elements.

Step-index fiber: A fiber whose refractive index of the core is uniformly higher than that of the surrounding cladding.

Survivability: A property of a system or device that provides some degree of assurance that the system or device will continue to operate despite a natural or artificial disturbance.

Synchronous: A transmission method in which timing signals at the sending and receiving stations control the synchronization of the data characters.

TCP/IP: Transmission control protocol/Internet protocol; protocol suite developed for the Internet. TCP is the primary transport protocol, while IP is the network layer protocol.

TDM: Time division multiplexing; a digital technique that allows information from multiple channels to be transmitted on a single link by assigning each channel a different time slot.

TDMA: Time division multiple access; a digital transmission technique for wireless communications systems that allows many users simultaneous access to a single radio frequency band without interference.

Throughput: Rate at which data is transmitted across a network.

TIA: Telecommunications Industry Association.

Token: A special frame that allows a node to transmit data.

Topology: Arrangement of nodes in a network; e.g., ring, bus, star, or tree.

Transpondent: A reception-only satellite device for receiving radio frequency signals.

Transport layer: Layer 4 of the OSI reference model.

Trunking: A process that allows a mobile to be connected to any unused channel in a group of channels of an incoming or outgoing call.

Twisted pair: A communication medium consisting of two copper wires twisted to minimize interference.

UHF: Ultra-high frequency; radio frequencies between 300 MHz and 3 GHz.

Virtual circuit: A path between two nodes established at the beginning of transmission by a packet switching mechanism.

VPN: Virtual private network.

VSAT: Very small aperture terminal; the earth-based antenna used for satellite data communications.

WAN: Wide area network; a network spanning a large geographical area, possibly the entire globe.

WDM: Wavelength division multiplexing; a transmission technique for assigning different wavelengths (colors) to different signals and passing them through a single optical fiber. It increases the capacity of a link without the need to install more fiber or to use higher speed transmission devices.

WER: Word error rate; the ratio of words received in error to the total number of words sent.

WLL: Wireless local loop; a fixed wireless system using radio technology such as cellular or PCS.

WPAN: Wireless personal area network.

Appendix — Physical constants

Electron charge	$e = 1.6 \times 10^{-19}$ C
Electron mass	$m = 9.1 \times 10^{-31}$ kg
Boltzmann's constant	$k = 1.38 \times 10^{-23}$ J/K
Permittivity of free space	$\varepsilon_o = 8.854 \times 10^{-12}$ F/m
Permeability of free space	$\mu_o = 4\pi x \times 10^{-7}$ H/m
Planck's constant	$\hbar = 6.6 \times 10^{-34}$ Js
Velocity of light in vacuum	$c = 3.00 \times 10^{8}$ m/s

Index

N

O

P

Q

R

S

T

U

V

W